MEDIA AND COMMUNICATIONS - TECHNOLOGIES, POLICIES AND CHALLENGES

ASSESSING THE DIGITAL NATION

BROADBAND ADOPTION, DEPLOYMENT AND USE

Media and Communications - Technologies, Policies and Challenge

Additional books in this series can be found on Nova's website under the Series tab.

Additional E-books in this series can be found on Nova's website under the E-book tab.

Internet Policies and Issues

Additional books in this series can be found on Nova's website under the Series tab.

Additional E-books in this series can be found on Nova's website under the E-book tab.

MEDIA AND COMMUNICATIONS -
TECHNOLOGIES, POLICIES AND CHALLENGES

ASSESSING THE DIGITAL NATION

BROADBAND ADOPTION, DEPLOYMENT AND USE

DARIUS L. SHEA
EDITOR

nova publishers
New York

For permission to use material from this book please contact us:
Telephone 631-231-7269; Fax 631-231-8175
Web Site: http://www.novapublishers.com

Additional color graphics may be available in the e-book version of this book.

Library of Congress Cataloging-in-Publication Data

ISBN 978-1-62948-362-7

Published by Nova Science Publishers, Inc. † New York

CONTENTS

PREFACE

Researchers and policymakers recognize that availability and use of high-speed Internet services – a range of connection technologies collectively known as broadband – are essential to economic growth. The United States has made considerable progress towards ubiquitous broadband access, with more than 93 percent of the population living in areas offering wired broadband service, and about 98 percent having access to either wired or terrestrial wireless connectivity at speeds of at least 3 Mbps download and 768 Kbps upload. The Internet has rapidly grown from an academic network into a resource that is now integral to the lives of most Americans. In 2000, only 4 percent of homes used broadband Internet service. By 2011, 69 percent were online at speeds greater than dial-up. By connecting every corner of our country to the digital age, we can help our businesses become more competitive, our students become more informed and our citizens become more engaged.

Additionally, our nation's technology industry leads the way in revolutionizing the nature of Internet use through mobile devices. Smartphones – driven by American-made operating systems and applications – now lead the mobile market. Over one billion smartphone users worldwide carry the global network in their pockets, including the 46 percent of Americans currently using these devices – now surpassing the 41 percent who use traditional mobile phones. Mobile broadband is vital to our future. This book examines online activities, broadband adoption, and reasons some do not go online at home which provide key factual underpinnings for the development of sound policies for increasing broadband adoption in the United States.

Chapter 1 - This report, based on data from the Census Bureau's July 2011 Current Population Survey ("CPS") Computer and Internet Use Supplement, updates and substantially expands on the previous study in this series, *Exploring the Digital Nation: Computer and Internet Use at Home*, issued in 2011. For the first time in eight years, this study goes beyond new findings on computer and Internet adoption in the United States to also present data on Americans' online activities – what Americans do once they connect to the Internet. The dataset contains information collected from more than 53,000 households, reporting on what Americans do online – including enjoying entertainment, communicating with friends, job searching and training, and researching health conditions and care, among other activities.

Below are highlights of the July 2011 CPS results. These findings about online activities, broadband adoption, and reasons some do not go online at home provide key factual underpinnings for the development of sound policies for increasing broadband adoption in the United States.

Chapter 2 - Broadband access is an essential part of our economy. Given the reliance of so many American businesses and families on this basic technology, it's easy to overlook that just 15 years ago broadband barely existed for consumers.

To create jobs and grow wages at home, and to compete in the global information economy, the delivery of fast, affordable and reliable broadband service to all corners of the United States must be a national imperative. The build-out of broadband infrastructure itself is a major driver of American investment and job creation, but even more significant are the ways that connectivity is transforming a range of industries, from education to entertainment to agriculture to travel. High-speed wired and wireless networks place the United States at the center of a digital economy that is one of the brightest parts of our short-term recovery and long-term competitiveness.

This report highlights (1) the status of broadband development in the United States in 2013, including both progress made in achieving greater availability to broadband as well as the significant remaining challenges of timely and affordable deployment of broadband to all Americans; (2) several important contributions of the Obama Administration over the last four years that have accelerated the growth of broadband technology; and (3) two key opportunities for further action.

Chapter 3 - On March 16, 2010, the Federal Communications Commission (FCC) released *Connecting America: The National Broadband Plan*. The National Broadband Plan (NBP) identified significant gaps in broadband availability and adoption in the United States, and in order to address those gaps and other challenges, the NBP set specific goals to be achieved by the year 2020. Goals were set for next generation broadband service; universal broadband service; mobile wireless broadband innovation and coverage; broadband access of Community Anchor Institutions; a nationwide, wireless, interoperable broadband public safety network; and broadband for tracking energy consumption.

Three years after the rollout of the NBP, available data indicate that there has been progress towards reaching the 2020 goals. The following observations can be made:

- the United States is much closer to reaching broadband availability goals than broadband adoption goals, which remain a major challenge;
- the United States is much closer to achieving broadband download speed goals than upload speed goals;
- while the next generation broadband goal of 100 million households with 100 Mbps speeds seems within reach (at least for download speeds), the price remains high—affordability could improve in the future depending on technological advances and consumer demand for ultra-high speed next generation performance;
- recent rollouts of next generation wireless technologies have led the FCC to state that the United States leads the world in mobile innovation; on the other hand, the latest Organisation for Economic Co-operation and Development (OECD) data indicate that the United States remains in the middle of the pack with respect to wireless broadband subscriptions per 100 of the population;
- while broadband data are incomplete for Community Anchor Institutions, available information indicate that the number of CAIs with 1 gigabit connections remains relatively low; and
- two major initiatives—FirstNet and Smart Grid—are currently underway in order to help reach the goals for a public safety wireless network and for broadband monitoring of energy consumption.

Many of the key telecommunications issues that are currently being considered by the 113th Congress are focused on improving broadband deployment and thus are intended to have a positive impact on the nation's progress towards reaching one (or in many cases, several) of the NBP goals. As the 113th Congress considers contentious telecommunications issues such as universal service reform, wireless technology and spectrum policy, and telecommunications regulatory reform, the ongoing progress towards meeting the NBP goals is likely to be part of that debate.

Chapter 4 - The "digital divide" is a term that has been used to characterize a gap between "information haves and have-nots," or in other words, between those Americans who use or have access to telecommunications and information technologies and those who do not. One important subset of the digital divide debate concerns high-speed Internet access and advanced telecommunications services, also known as broadband. Broadband is provided by a series of technologies (e.g., cable, telephone wire, fiber, satellite, wireless) that give users the ability to send and receive data at volumes and speeds far greater than traditional "dial-up" Internet access over telephone lines.

Broadband technologies are currently being deployed primarily by the private sector throughout the United States. While the numbers of new broadband subscribers continue to grow, studies and data suggest that the rate of broadband deployment in urban/suburban and high income areas is outpacing deployment in rural and low-income areas. Some policymakers, believing that disparities in broadband access across American society could have adverse economic and social consequences on those left behind, assert that the federal government should play a more active role to avoid a "digital divide" in broadband access.

With the conclusion of the grant and loan awards established by the American Recovery and Reinvestment Act of 2009 (P.L. 111-5), there remain two ongoing federal vehicles which direct federal money to fund broadband infrastructure: the broadband and telecommunications programs at the Rural Utilities Service (RUS) of the U.S. Department of Agriculture and the Universal Service Fund (USF) programs under the Federal Communications Commission (FCC). Although the USF's High Cost Program does not explicitly fund broadband infrastructure, subsidies are used, in many cases, to upgrade existing telephone networks so that they are capable of delivering high-speed services. Additionally, subsidies provided by USF's Schools and Libraries Program and Rural Health Care Program are used for a variety of telecommunications services, including broadband access. Currently the USF is undergoing a major transition to the Connect America Fund, which is targeted to the deployment, adoption, and utilization of both fixed and mobile broadband.

To the extent that the 113th Congress may consider various options for further encouraging broadband deployment and adoption, a key issue is how to strike a balance between providing federal assistance for unserved and underserved areas where the private sector may not be providing acceptable levels of broadband service, while at the same time minimizing any deleterious effects that government intervention in the marketplace may have on competition and private sector investment.

In: Assessing the Digital Nation
Editor: Darius L. Shea

ISBN: 978-1-62948-362-7

Chapter 1

EXPLORING THE DIGITAL NATION: AMERICA'S EMERGING ONLINE EXPERIENCE*

U.S. Department of Commerce

FOREWORD

Researchers and policymakers recognize that availability and use of high-speed Internet services – a range of connection technologies collectively known as broadband – are essential to economic growth. The United States has made considerable progress towards ubiquitous broadband access, with more than 93 percent of the population living in areas offering wired broadband service, and about 98 percent having access to either wired or terrestrial wireless connectivity at speeds of at least 3 Mbps download and 768 Kbps upload (NTIA & FCC, 2013). The Internet has rapidly grown from an academic network into a resource that is now integral to the lives of most Americans. In 2000, only 4 percent of homes used broadband Internet service. By 2011, 69 percent were online at speeds greater than dial-up. The President has observed, however, that we have more work to do.

* This is an edited, reformatted and augmented version of the National Telecommunications and Information Administration and Economics and Statistics Administration, dated June 2013.

By connecting every corner of our country to the digital age, we can help our businesses become more competitive, our students become more informed and our citizens become more engaged.

—President Obama, announcing U.S. Ignite, June 12, 2012

Additionally, our nation's technology industry leads the way in revolutionizing the nature of Internet use through mobile devices. Smartphones – driven by American-made operating systems and applications – now lead the mobile market. Over one billion smartphone users worldwide carry the global network in their pockets, including the 46 percent of Americans currently using these devices – now surpassing the 41 percent who use traditional mobile phones (Pew Internet, 2012). Mobile broadband is vital to our future.

For our families and our businesses, high-speed wireless service, that's the next train station; it's the next off-ramp. It's how we'll spark new innovation, new investment, new jobs.

—President Obama, addressing Northern Michigan University, February 10, 2011

One of the Obama Administration's priorities is to increase broadband use in the United States, which is a key ingredient for job creation and sustainable economic growth. The National Telecommunications and Information Administration ("NTIA") has been directly implementing this policy in a range of ways, including:

- Overseeing nearly $4 billion of Broadband Technology Opportunities Program ("BTOP") and State Broadband Initiative ("SBI") grants;
- Working to reallocate 500 MHz of spectrum for commercial wireless broadband service;
- Developing and maintaining the National Broadband Map;
- Creating a digital literacy portal with the Department of Education and eight other agencies; and,
- Analyzing broadband availability and adoption data in collaboration with the Economics and Statistics Administration ("ESA") of the Department of Commerce.

The Administration strongly believes that good data begets sound policymaking, and to that end, the Department of Commerce ("the

Department") is the leading source of published data on broadband availability and Internet use in America. The Department collects data from several sources, including BTOP, SBI, and the Computer and Internet Use Supplement to the Census Bureau's Current Population Survey ("CPS"). These datasets are publicly available for use by researchers to conduct economic, financial, demographic, and other studies.

The National Broadband Map, which is a searchable database that includes over 125 million records, has already attracted nearly one million unique users and served over 100 million third-party data requests. Unveiled in February 2011 and updated every six months, the map offers a research tool that details individual Internet service providers, showing where they offer service, the maximum speeds they advertise, and how much of a geographic area – down to the Census block level – they cover.

While the map provides extensive data on the availability of broadband, the CPS Computer and Internet Use Supplement collects detailed data about Internet adoption – who actually goes online, and how. While over 90 percent of Americans live in areas where high-speed Internet is available, only seven in ten households used broadband at home by July 2011. Why the disparity? Overall, 48 percent of households lacking home Internet connections reported that they had no need to use the Internet at home, or were uninterested in it. Affordability is the next largest concern, cited by 28 percent of non-using households. The lack of an adequate (or perhaps any) computer ranked third (13 percent) among reasons given by households not online.

At NTIA's request, the July 2011 CPS Computer and Internet Use Supplement included a significant new series of questions regarding online activities. The answers to these questions provide valuable insights into how people use the Internet. For example, Americans ages 25 and older frequently reported relying on the Internet for personal communications, information, financial services, and other activities. One of the most significant findings is that once disadvantaged groups overcome the hurdles of getting online, they – more than other users – conduct job searches online. This helps not only the individual but also the nation's economy.

The CPS Supplement reviewed in this report is the first large-scale data collection in eight years of how people use the Internet, and should be invaluable for further study. With these and future data, we seek to monitor America's embrace of the digital age and accelerate our progress in this regard through effective policymaking.

EXECUTIVE SUMMARY

This report, based on data from the Census Bureau's July 2011 Current Population Survey ("CPS") Computer and Internet Use Supplement, updates and substantially expands on the previous study in this series, *Exploring the Digital Nation: Computer and Internet Use at Home*, issued in 2011. For the first time in eight years, this study goes beyond new findings on computer and Internet adoption in the United States to also present data on Americans' online activities – what Americans do once they connect to the Internet. The dataset contains information collected from more than 53,000 households, reporting on what Americans do online – including enjoying entertainment, communicating with friends, job searching and training, and researching health conditions and care, among other activities.

Below are highlights of the July 2011 CPS results. These findings about online activities, broadband adoption, and reasons some do not go online at home provide key factual underpinnings for the development of sound policies for increasing broadband adoption in the United States.[1]

WHAT AMERICANS DO ONLINE

The July 2011 CPS data collection gathered information on a broad range of online activities.

This report focuses primarily on three areas – employment, health, and civic engagement – that are particularly important for society. The CPS data suggest that widespread Internet use benefits society, that mobile devices further increase these benefits, and that the Internet's great utility leads users to go online regularly and rely on it in their daily lives.

- **Employment:** Seventy-three percent of unemployed Internet users reported going online to look for work, as did 52 percent of underemployed users. Online career training appeared to be most popular among those Americans who are currently employed.
- **Health:** Researching health plans and finding medical information were relatively common online activities. Less common, however, was online interaction with healthcare professionals, and telemedicine remains in its infancy.

- **Civic Engagement:** The Internet may be significantly more effective than television – and closer to traditional print media and radio – in encouraging civic engagement, particularly in light of the Internet's emergence as a news medium.
- **Other Activities:** Large majorities of users relied on the Internet for personal communications (77 percent) and general information (66 percent). The Internet was also a popular venue for financial services, as 53 percent of users relied on the Internet for banking, and 52 percent relied on it for shopping and other consumer services. Additionally, nearly half of users depended on the Internet for entertainment, and a third relied on it for on-the-go services such as real-time directions. Seventy-two percent of Internet users went online daily for these and a broad range of other activities.

How Americans Get Online

In considering how Americans connect to the Internet, the data show that both mobility and higher-speed connections are increasingly becoming hallmarks of daily life. Reliability and cost are among the most important factors for households in choosing an Internet provider. Disparities in broadband adoption based on demographics and geography persist, but have generally diminished over time.

- **Mobile Broadband is Growing:** While personal computers were still the preferred means of accessing the Internet, mobile phones were a strong complement to personal computers. Thirty-nine percent of people who accessed the Internet from home used both a personal computer and a mobile device. Tablets were not a primary means of going online for most Internet users in 2011, with only 9 percent saying they used a tablet to go online.
- **Broadband Technologies are Dominant:** The types of technology that households have adopted to access the Internet have changed over time. Dial-up use has steadily declined, from 11 percent of households in 2007 to just 2 percent in 2011, so that virtually all households that used the Internet at home in 2011 did so via broadband technologies. Twenty-eight percent of households did not use the Internet at home, but 69 percent of households had some type of broadband Internet

service at home, compared to only half (51 percent) of households in 2007.

- **Service Reliability is a Key Factor:** Among households using the Internet at home, 37 percent indicated they considered reliable service the most important factor in determining their choice of Internet service provider, closely followed by connection speed (33 percent), and then affordability (24 percent).
- **Cost is Also Important:** Thirty-eight percent of households that switched Internet service providers cited differences in cost as the main reason for doing so, followed by 30 percent of households that changed providers in order to obtain faster connection speeds. Thirty-six percent of households with broadband Internet service reported they paid between $30 and $44 per month for non-bundled service. Only 6 percent of households with broadband reported paying less than $15 per month, while 16 percent reported paying $60 or more per month.

THE WHO AND WHERE OF GOING ONLINE

The prevalence of Internet use differs by demographic and geographic characteristics. In aggregate, seven out of ten households and about two out of three persons ages 16 and older used broadband at home by 2011. However, low-income, non-Asian minority, and rural households were much less likely to be connected than their more affluent, urban, and white or Asian American counterparts. Additionally, Internet users frequently go online at locations outside the home, including many who do not live in households with Internet connections. Disaggregating data in this fashion can help researchers and policymakers to understand better the nature of Internet use in the United States.

- **Impact of Demographics:** The propensity for households to own computers, use the Internet, and adopt broadband varied based on demographic characteristics. Low-income and less educated households experienced computer ownership and broadband adoption rates well below the national average. Furthermore, lower percentages of African American and Hispanic households adopted broadband Internet in 2011, compared to all households and to white and Asian American households. Households headed by someone with a disability also had lower levels of computer and Internet use. Households with school-age children were more likely than those without children to own a computer and to adopt broadband. Younger households had higher rates of computer ownership and broadband adoption compared to households headed by a senior citizen.

- **Urban-Rural Divide:** Geographic location was also associated with differing adoption rates. Urban households had higher rates of computer ownership and broadband adoption than their rural counterparts. Compared to the national adoption rate, white and Asian American households, and high-income and highly educated households, had higher computer ownership and broadband adoption rates in both urban and rural areas. Rural African American and Hispanic households with low incomes reported the lowest computer and broadband adoption rates.
- **State-by-State Data:** At the state level, the proportion of households with a computer at home ranged from 65 percent to 85 percent. State broadband adoption rates ranged from 53 percent to 80 percent. Broadband adoption rates in states tended to be higher in urban locations than in rural locations, with the size of the urban-rural gap ranging from 1 percentage point to 29 percentage points in 2011.
- **Place of Access:** Ninety-two percent of people who accessed the Internet did so from home. Internet users who did not have a computer at home went online at public libraries, work, school, and a range of other places including other people's houses, cafés, and community centers.

WHY SOME AMERICANS DO NOT GO ONLINE AT HOME

A key ongoing focus of the CPS data collections in the *Digital Nation* series is to look closely at households that do not have broadband Internet service to assess why not. Despite the tendency most have developed for accessing the Internet from home, approximately 30 percent of the 119 million households represented in the CPS did not use the Internet at home, which contributed to the persistence of the digital divide. While that gap of digitally disconnected households has continued to shrink in recent years, households reported three primary reasons for not using the Internet where they live: They do not need or are not interested, the Internet is too expensive, or they lack adequate computing equipment.

- **Big Picture:** Almost half of households not using the Internet at home (48 percent) stated they do not need the Internet or are not interested in accessing this important resource. Another 28 percent of households without Internet at home cited expense as the main deterrent to their use, and 13 percent explained they did not have a computer at all or one that worked well enough to go online.
- **Lack of Interest or Need:** Households whose members have never used the Internet in their homes were most likely to say they had no interest in or need for such service at 52 percent, compared to 20 percent of former home Internet users who no longer had such service. Similarly, 52 percent of households without children under 18 living at home expressed disinterest in home Internet use, making them twice as likely as households with school-age children at home (26 percent) to cite this reason. The high incidence of disinterest in home Internet use among those who have never used the Internet at home and those without school-age children living with them may reflect these groups' lack of exposure to the benefits and relative ease of Internet use that experienced Internet users in the household could provide.
- **Too Expensive:** The percentage of households reporting that home Internet service was "too expensive" increased by 4 points in the nine-month period between the 2010 and 2011 data collections, possibly reflecting some consumers' concern about their personal financial circumstances during a period of slow economic recovery. Forty-one percent of former users discontinued home Internet service because of its expense.
- **No Adequate Computer:** Although the number of households without a computer or one adequate enough to connect to the Internet declined by 2 percentage points between 2010 and 2011 to 13 percent, 57 percent of households citing this reason for not using the Internet in their homes earned less than $25,000 annually.
- **Lack of Availability:** Among non-using households, a lack of service availability was not one of the key obstacles to using the Internet at home. For example, only 2 percent of rural households cited no Internet service in their area as their primary reason for not using the Internet at home, compared to 1 percent of urban households. However, the data still show a significant unmet demand for broadband among dial-up Internet users. Twenty-one percent of dial-

up users stated they did not use broadband at home because it was not available in their area.

The July 2011 CPS Computer and Internet Use Supplement, which surveyed more than 53,000 households, is a critical source of data about how Americans use the Internet, how they go online, and why some do not. This report highlights some of the key findings from the data, and lays a foundation for a broad array of academic and policy researchers to explore the dataset and reach additional conclusions to inform policy decisions about broadband and the Internet.

1. Introduction: Broadband Internet in the United States

The Internet plays a vital role in most Americans' lives, and is increasingly becoming a key tool for job searches and training, for research on health issues, and for active participation in our society. Since its commercialization in the 1990s, the Internet has brought sweeping changes in the ways Americans communicate, gather information, conduct commerce, and entertain themselves. Widespread Internet use provides numerous societal benefits, including increased civic engagement,[2] economic growth, and enhanced productivity (NTIA, 2011, p. 6).

To inform public policy decisions that maximize the benefits of a connected society, the Department of Commerce has been at the forefront of gathering critical information about the availability, adoption, and use of Internet services for almost twenty years. The National Telecommunications and Information Administration ("NTIA") sponsors the Computer and Internet Use Supplement to the Census Bureau's Current Population Survey ("CPS") to gather the data on Internet adoption and usage habits that form the basis for this report. This report – prepared by NTIA and the Economics and Statistics Administration ("ESA") – provides highlights from the latest of this series of data collections, and for the first time in eight years, the data include an array of information on how Americans use broadband services.

As the National Broadband Map demonstrates, high-speed Internet is widely available in the United States. The map shows that 98 percent of Americans live in areas where broadband Internet connections are available with download speeds of at least 3 Mbps and upload speeds of at least 768 Kbps, through either wired or wireless technologies (see Table 1).[3]

In sharp contrast to the near ubiquitous availability of high-speed Internet, actual Internet use is not yet universal. Data in the 2011 CPS Supplement show that household use of computers, the Internet, and specifically broadband Internet have all increased significantly during the last two decades (see Figure 1).[4] Between 2000 and 2011, household computer use increased by 49 percent, while home Internet use rose by 71 percent, and broadband Internet use soared by over 1,600 percent. For comparison, household use of traditional telephone services remained relatively constant, rising by just 2 percent during that period (FCC, 2011).

Despite this growth, Internet adoption rates lag behind availability in the United States. Approximately 69 percent of households used broadband Internet at home (72 percent if including dial-up) in July 2011 (see Figure 1).[5]

Nationwide results mask the more complex details about who does and does not go online. Internet use varies significantly among different demographic groups and geographic areas.[7] In the United States, factors such as income, educational attainment, age, gender, race, employment status, and disability status correlate with uneven Internet adoption rates. Geography also makes a difference, including population density and state. Additionally, those households that do not go online at home reported a range of underlying reasons for non-use. Lack of need or interest and affordability ranked as the major reasons for non-use at home. However, the reasons varied in importance depending on the demographics or geography involved.

Beyond the detailed data on Internet adoption, the 2011 CPS Supplement included questions about online activities. This significantly expanded survey contained new questions asking for which daily activities Americans used and relied on the Internet. ESA and NTIA analysis of these results suggest that the Internet has become an important tool for Americans seeking to advance their careers, learn about health care options, and become active citizens in their communities. Furthermore, Internet users frequently reported that they rely on going online for personal communications, general information, financial services, consumer services, and entertainment. These newly collected data also reveal, among other findings, that mobile devices facilitate the use of a wide range of online applications. This comes at a time when half of all mobile phone users in the United States use smartphones (The Nielsen Company, 2012).

Table 1. Nationwide Availability of Broadband by Minimum Speed and Technology Type, Percent of Household Units and Population, 2011

Technology	Download Speed: 3 Mbps Upload Speed: 768 Kbps			
	Household Units	% Margin of Error (+ / -)	Population	% Margin of Error (+ / -)
Any Technology	97.7	0.7	98.2	0.6
Wireline	92.8	0.0	93.4	0.0
Wireless	92.8	2.0	93.9	1.8

Source: National Broadband Map, Broadband Statistics Report, "Access to Broadband Technology by Speed" (NTIA & FCC, 2013).

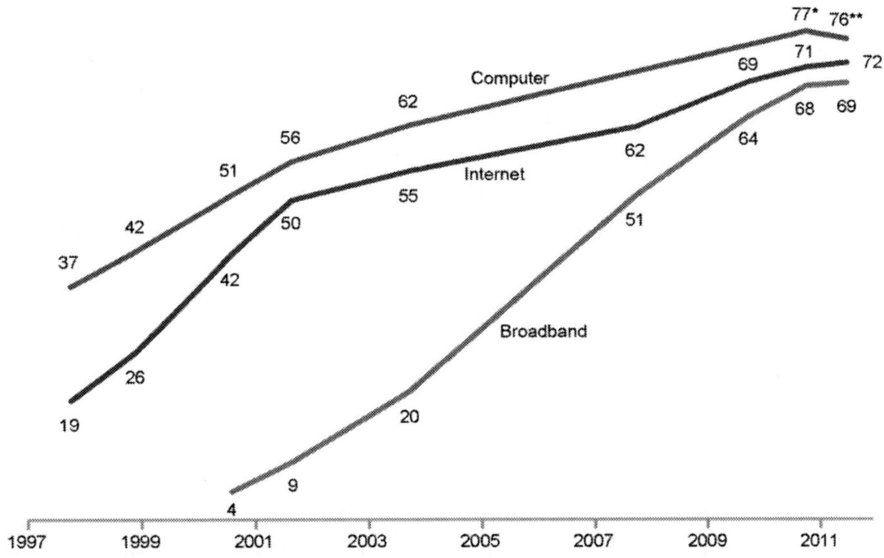

* Includes handheld devices, such as smartphones and tablets

** Includes tablets but not smartphones

Figure 1. Overview of Household Adoption Rates by Technology, Percent of U.S. Households,[6] 1997-2011.

The sections that follow probe various dimensions of Internet use in the United States. Section 2 discusses what people reported doing while online, focusing particularly on the societal benefits of specific activities, the effects of mobile devices on usage patterns, and the extent to which the country relies on the Internet and the applications it enables. Section 3 details the devices and Internet technologies Americans used to go online, the factors that motivated them to use the Internet and switch providers, the costs of Internet service, and demographic disparities in Internet use. Non-adoption of the Internet is the focus of Section 4, including the main reasons households gave for not having Internet connectivity at home. This study focuses particular attention on demographic and geographic differences in usage patterns and the implications for the country.

2. Online Activities: Why Connectivity Matters

Since NTIA published its first report on computer and Internet use nearly two decades ago, this series of studies has focused primarily on adoption rates and the digital divide between various demographic and geographic groups. Universal Internet use in the United States is a critical national policy objective, and over the years ESA and NTIA have examined gaps in adoption to understand better how to eliminate them. But *why* does Internet use matter, and what makes adoption by all Americans an appropriate goal for policymakers?

Extensive external research underlies the Administration's belief in the great value of universal Internet adoption to the nation. Academic studies suggest that Internet use increases employment and income, enhances consumer welfare, and promotes civic engagement (Brynjolfsson, Hu, & Smith, 2003; Crandall, Lehr, & Litan, 2007; Gillett, Lehr, Osorio, & Sirbu, 2006; Stenberg et al., 2009; Tolbert & McNeal, 2003). Using the Internet may even reduce damage to the environment and improve access to health information (Choo, Mokhtarian, & Salomon, 2005; Rains, 2008; Romm, 2002). In making information exchange simple, fast, and constant, the Internet enables users to lead lives that are more informed, efficient, and fulfilling.

In support of the academic community, and in an effort to expand the scope of the Department's own research to include the benefits of Internet use, NTIA worked with the Census Bureau to expand substantially the scope of the CPS Computer and Internet Use Supplement's July 2011 data collection. For the first time since 2003, the CPS Supplement included extensive questions about the ways in which people use the Internet, including the applications they utilize, their devices of choice, and the aspects of life for which they rely on the global network. The data show how often Americans went online, the places in which they did so, and the relationship between socioeconomic factors and these actions. The collected data – highlighted below – suggest that widespread Internet use benefits society, that mobile devices further increase these benefits, and that the Internet's great utility leads users to go online regularly and rely on it in their daily lives.

Societal Benefits

The CPS Supplement results, taken together with the extensive body of literature on the benefits of Internet use, suggest that increasing adoption can benefit society by boosting employment, improving public health, and facilitating active citizenship. The open and egalitarian nature of the Internet enables developers to create an impressive range of applications, limited primarily by the imagination. These applications, created in large part by American innovators, enable face-to-face conversations with friends and family, provide maps and directions to unfamiliar places, and support instantaneous searches through entire libraries of human knowledge. Most Americans have clearly determined the Internet to be worth the cost of service and equipment, as demonstrated by its rapid adoption during the first twenty years of public availability. Yet the CPS Supplement data confirm that Internet use remained far from universal in the United States in 2011. If increased adoption benefits society in addition to the users themselves, then the marketplace for Internet access may not reach its optimal size without additional investment in infrastructure, digital literacy training, or other forms of support. The likelihood that society may indeed benefit from widespread Internet use is examined below.

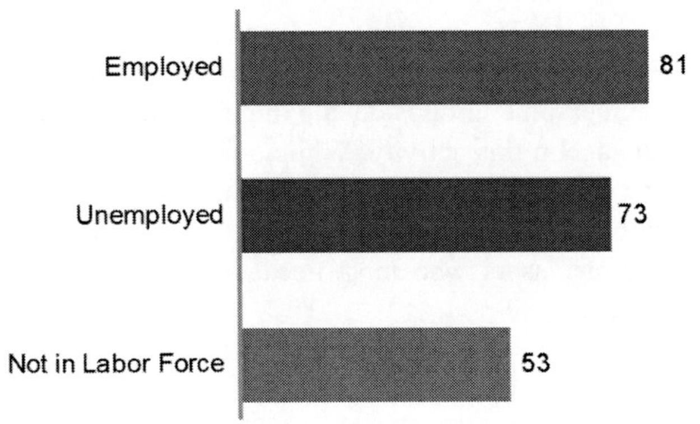

Figure 2. Internet Use by Employment Status, Percent of Persons Age 25+, 2011.

Employment

Given the direct correlation between income and Internet use established in previous *Digital Nation* reports (ESA & NTIA, 2011), it is not surprising that employed Americans were more likely to go online than their peers who were not working. Among the nationwide population ages 25 and older, 81 percent of employed persons were Internet users in July 2011, compared with 73 percent of the unemployed[8] and just 53 percent of those persons not in the labor force (see Figure 2).[9] Even among employed Americans, the underemployed (those working fewer than 35 hours per week who want to work full-time) faced a similar disadvantage compared with fully employed persons.[10] Only 71 percent of the underemployed were Internet users, compared with 81 percent of fully employed persons. This disparity prompts examination of whether decreased Internet use is merely a symptom of diminished income or other factors, or if evidence suggests a correlation between Internet use and employment outcomes.

One of the ways in which Internet use may affect employment is through the ability to search and apply for jobs online, especially in light of the fact that many leading employers, including the Federal Government, require online job applications.[11] Looking for work was a relatively frequent activity for Americans using the Internet, with 34 percent of Internet users reporting that they go online when conducting a job search. Unemployed Internet users were nearly twice as likely to look for work online as their employed counterparts, with 73 percent of the unemployed using the Internet for this activity. For comparison, 37 percent of employed Internet users conducted job searches online, and only 18 percent of those not in the labor force reported this activity (see Figure 3). Similarly, 52 percent of underemployed persons who went online used the Internet to look for jobs, compared with 36 percent of fully employed users. The apparent tendency for Internet users to go online for job searching when they are in immediate need of employment may help explain why demographic groups facing higher unemployment rates were more likely to engage in this activity. While only 32 percent of white Internet users went online when conducting a job search, 35 percent of Hispanic users and 41 percent of African American users reported this activity.[12] Across communities, Internet users who most needed to find a job went online to advance their search.

While searching for jobs online was most frequent among those Internet users actively seeking employment, Internet-based job training was somewhat more common among those who were currently working. Twenty-eight percent of employed Internet users reported going online for job training or courses, while 25 percent of their unemployed counterparts participated in this activity. Interestingly, underemployed Internet users were less likely (21 percent) to participate in job training online than other employed users (28 percent), suggesting that such training may be more accessible to those persons with a full-time employer.

> NTIA and other federal partners developed **DigitalLiteracy.gov** as a valuable resource for practitioners providing digital literacy training and services in their communities. The portal includes best practices that grantees are implementing in their projects, and other useful tools. NTIA collaborated with the U.S. Departments of Education, Agriculture, Energy, Health and Human Services, Housing and Urban Development, and Labor, as well as the Corporation for National and Community Service, the Federal Communications Commission, and the Institute of Museum and Library Sciences. The portal offers online resources on many topics, including workforce training and online job searching.

Education and family income levels also reveal disparities in Internet use for job training or courses. Only 9 percent of Internet users without high school diplomas participated in online job training or courses, compared with 12 percent of high school graduates, 23 percent of those with some college credit, and 32 percent of college graduates (see Figure 4). Similarly, 15 percent of online Americans with annual family incomes below $25,000 used the Internet for job training, compared with 31 percent of those with family incomes of $100,000 or more.

Americans clearly find the Internet useful in supporting their career prospects. Job seekers frequently go online to search for opportunities, while employed persons may use the Internet to further develop their skills. These usage patterns and existing research prompt examination of whether Internet use is positively correlated with employment, holding constant other factors believed to influence employment including age, education, gender, marital status, race, disability, household size, population density, and region. The large volume of regularly-asked CPS questions on employment, which the Bureau of Labor Statistics ("BLS") uses to calculate the official labor force estimates, helps in introducing evidence on this question. Multivariate regression analysis enables estimation of the extent to which Internet use and the factors mentioned above are correlated with an individual's probability of employment. Appendix B details the model used for this analysis (see Table B1).

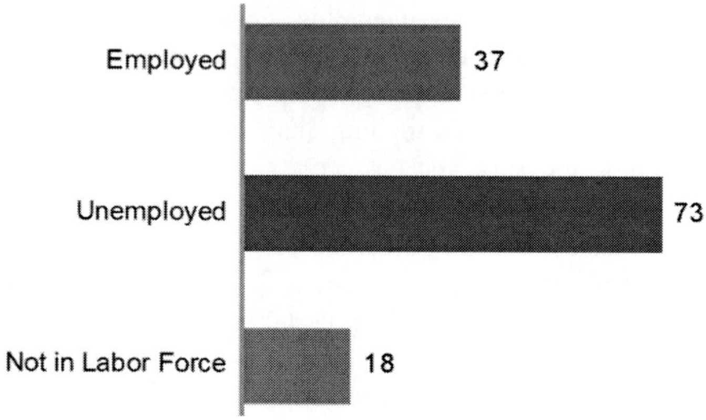

Figure 3. Online Job Searching by Employment Status, Percent of Internet Users Age 25+, 2011.

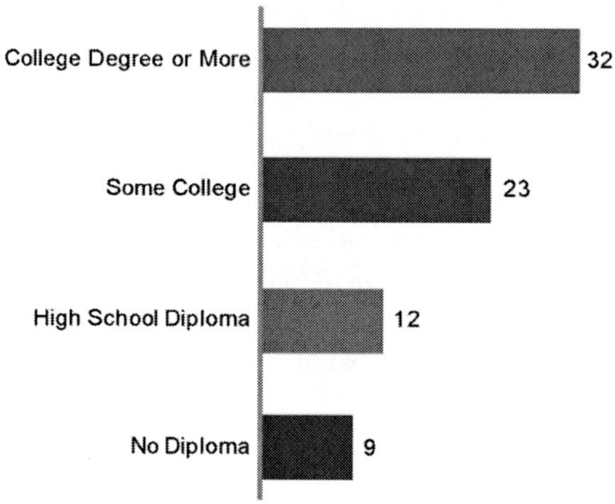

Figure 4. Job Training Online by Education, Percent of Internet Users Age 25+, 2011.

After controlling for demographic and geographic factors, Internet use among adults ages 25 and older was associated with a 6 percentage point increase in probability of employment, relative to individuals who were not online. Furthermore, living in a household where someone goes online from home – regardless of whether the individual personally does so or not – was associated with a 3 percentage point increase in the probability of having a job. The marginal effects are cumulative, so the model suggests that fitting into both of the aforementioned categories was correlated with a 9 percentage point increase in probability of employment, relative to individuals who neither used the Internet nor resided in a household where the Internet was used (see Figure 5). The model also shows that variables like age, gender, education, and race correlate with employment outcomes.[13] While this regression analysis alone cannot establish the extent to which Internet use might improve employment outcomes, any benefits that exist go beyond the obvious returns to individual Americans and their families. Increased employment rates improve the economy, boost tax revenue to support schools and other vital public services, and, according to researchers, may even lead to decreased crime (Bausman & Goe, 2004).

These findings alone do not prove, nor can any one study of survey data establish, that Internet use causes employment. As many jobs require utilizing the Internet for communication and other purposes, it is plausible that employment causes Internet use rather than the other way around, or that each factor encouages the other. Additionally, the mutual association with family income may confound the observed relationship between the two factors.

Employment leads to higher income, and income enables individuals to purchase the equipment and services required to go online. Given the existence of other studies suggesting positive effects of Internet use on employment (Crandall, Lehr, & Litan, 2007; Gillett, Lehr, Osorio, & Sirbu, 2006), these findings provide some additional support for the theory. Taken together with the reported findings on Internet use specifically for job searching and training, the Internet appears to be useful in improving the prospects for employment.

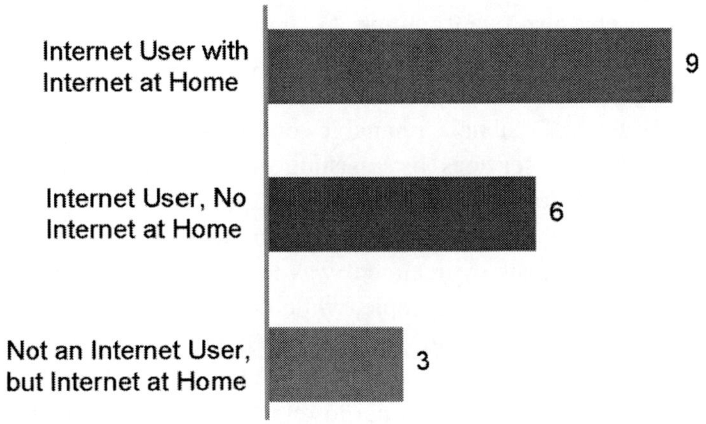

Figure 5. Marginal Effects of Internet Use on Employment, Percentage Point Increase for Persons Age 25+, 2011.

Healthcare

In addition to assisting Americans in advancing their careers, the Internet may also improve healthcare outcomes among its users. Going online to research information on health plans or practitioners, for example, is a relatively common activity, reported by 35 percent of Internet users. Given the Internet's considerable strength as an information source, using it for this purpose may enable individuals to make better choices about their healthcare options.[14] As with employment-centric tasks, use of the Internet for this type of healthcare research was significantly more common among some demographic groups than others. Thirty-seven percent of Internet users between ages 25 and 44 reported looking up health plans or practitioners online, and 35 percent of those between 45 and 64 did so. Yet only 29 percent of those users ages 65 and older researched health plans or practitioners online. The data also suggest significant disparities based on population density. Thirty-seven percent of Internet users living in urban[15] areas used the Internet to research health plans or practitioners, compared with only 26 percent of those living in rural areas. Differences in income and education also reveal gaps in utilizing the Internet for healthcare research. Only 19 percent of Internet users lacking a high school diploma went online for healthcare research, compared with 22 percent of high school graduates, 34 percent of those with some college credit, and 47 percent of college graduates (see Figure 6). Similarly, while 48 percent of users with annual family incomes of $100,000 or more used the Internet for this purpose, only 24 percent of those with incomes below $25,000 per year reported researching healthcare information online.

Internet users also went online to find health information for self-diagnosis or treatment, with 42 percent of users engaging in this activity. While finding medical information online is no substitute for consulting a trained medical professional, reputable online sources may complement traditional healthcare services by assisting Americans in properly treating minor ailments and determining when to seek professional medical attention. As with researching health plan and practitioner information, using the Internet for self-diagnosis or treatment was far more common among certain groups of Internet users. For example, while only 31 percent of Internet users with annual family incomes below $25,000 went online to research medical treatment, 52 percent of those with family incomes of $100,000 or more reported using the Internet for this purpose (see Figure 7). As is the case with other online activities, groups with differing education levels also exhibit similar disparities. Fifty-one percent of Internet users with college degrees went online to research health information for self-diagnosis or treatment, compared with 42 of users with some college credit, 30 percent of those with high school diplomas, and just 22 percent of those without a high school diploma. White and Asian American Internet users were more likely to engage in this online activity (44 percent and 40 percent, respectively) compared with African Americans (33 percent) and Hispanics (32 percent). Perhaps surprisingly, the rate at which Americans used the Internet for medical research did not vary dramatically based on age group; 42 percent of Internet users between 25 and 44, 42 percent between 45 and 64, and 38 percent ages 65 and older reported conducting this type of research online.

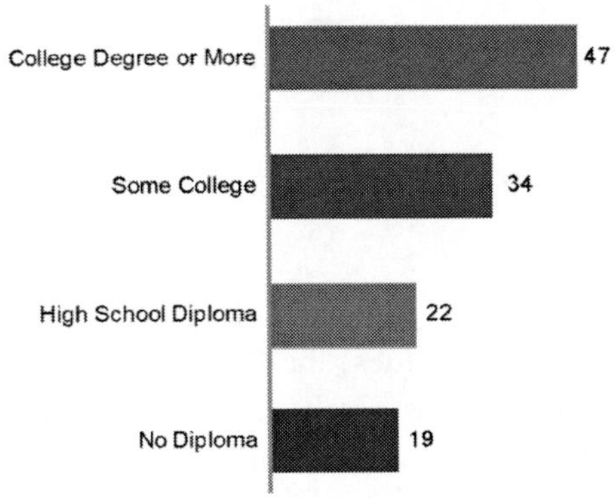

Figure 6. Researching Health Plan or Practitioner Information Online by Education, Percent of Internet Users Age 25+, 2011.

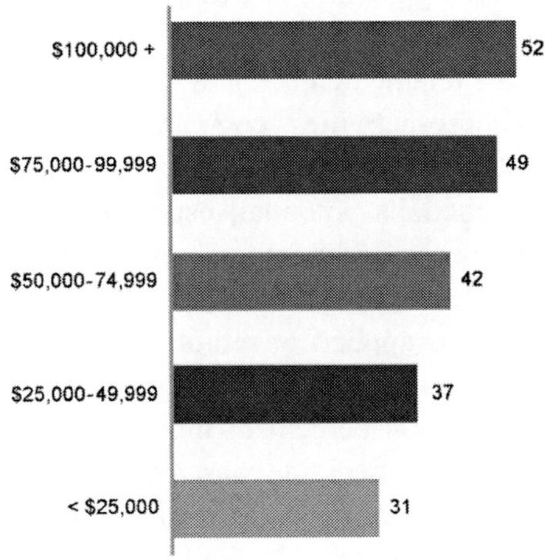

Figure 7. Researching Medical Treatment Information Online by Family Income, Percent of Internet Users Age 25+, 2011.

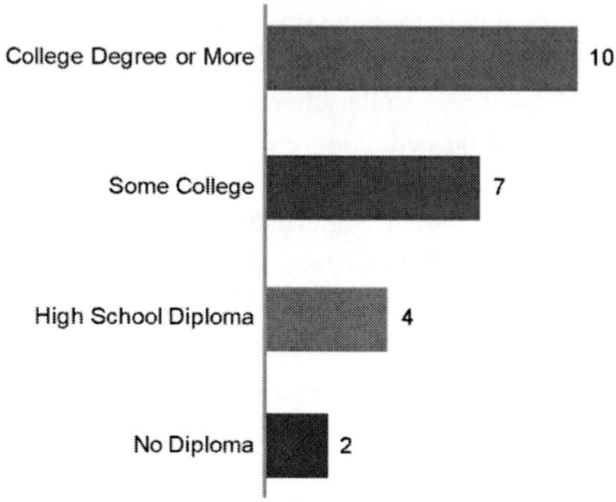

Figure 8. Using Telemedicine Online by Education, Percent of Internet Users Age 25+, 2011.

While researching health plans and finding medical information were relatively common online activities, the 2011 CPS Supplement data suggest that Internet users are far less likely to interact electronically with their healthcare professionals. Only 7 percent of Internet users reported going online to access medical records, participate in video conferencing with a doctor, or take advantage of remote procedures such as heart rate monitoring. Although these activities were not common among any demographic groups, a few interesting trends are apparent. For example, Internet users living in urban areas were twice as likely to participate in online healthcare activities (sometimes known as "telemedicine"[16]); 8 percent of users in urban areas reported engaging in these activities, compared with 4 percent of their rural counterparts. The relatively low uptake of telemedicine in rural areas suggests that it has not yet reached its potential, especially given that some experts consider telemedicine particularly important in rural areas where medical professionals may be miles away (Harrison & Lee, 2006). Disparities based on income and education also applied to remote healthcare activities. While 11 percent of Internet users with family incomes of $100,000 or more engaged in telemedicine activities, only 4 percent of those in the under $25,000 bracket reported such activities. Telemedicine was also much more common among users with college degrees, at 10 percent, than those with no high school diploma, at 2 percent (see Figure 8). Perhaps less predictably, few racial disparities were evident in utilization of telemedicine, although Asian American Internet users (11 percent) were significantly more likely to undertake such activities than whites (7 percent), African Americans, or Hispanics (each 6 percent). In contrast to the other two healthcare-related activities discussed above, telemedicine remains in its infancy.

Although it is beyond the scope of this report to attempt to estimate the Internet's effect on health policy challenges, it is clear that Internet users find utility in going online to get information about health insurance and providers, as well as to assist in self-diagnosis and treatment of health issues. Such activities may allow Internet users to leverage online information to make more-informed choices about their healthcare.

Civic Engagement

While the benefits of increased employment and improved healthcare are readily understood, the concept of civic engagement may be less familiar. For several decades, social scientists have used civic engagement to refer to a range of activities through which people participate in their communities. In his landmark study of social capital, Robert Putnam recognizes several forms of civic engagement, including political activity, family and neighborhood connections, and organizational membership (Putnam, 2000). Many believe that civic engagement boosts trust, cooperation between neighbors, and participation in the country's institutions. Given Putnam's observation that the American tradition of civic engagement is important to the nation's well-being, Internet applications that stimulate greater involvement in civic life and strengthen democracy merit further study.

As a communications and information tool, the Internet has great potential to influence civic engagement. For example, 56 percent of Internet users went online to obtain news or other information. In view of research suggesting that media attentiveness may help predict civic engagement (Jennings & Zeitner, 2003), it is encouraging that most Internet users found utility in going online to learn about current events. Although television remained the primary source of news for more than half of Americans who go online, 22 percent cited the Internet as their primary source (see Figure 9). And among Internet users who do go online for at least *some* news, about the same percentage chose the Internet as their primary source as cited television (39 percent relying on the Internet as their primary source versus 42 percent for television).

Although the Internet had yet to surpass television as America's primary news medium in 2011, certain groups of Internet users were more likely than others to report that going online was their primary means of learning about current events. Age is a particularly strong predictor of whether an Internet user reported this behavior, with younger individuals being far more likely to use the Internet as their primary news source. While 31 percent of Internet users between 25 and 44 did so, only 17 percent of those between 45 and 64 and 8 percent of those ages 65 and older primarily went online to get their news. Older Americans who used the Internet were more likely to turn primarily to television or print media for news or other information (see Figure 10). The survey results show significant racial differences in the Internet's prominence as a news source; one-third of Asian American Internet users primarily went online for news, compared with 23 percent of whites and 17 percent of both African Americans and Hispanics. Additionally, male Internet users were more likely than women to have listed the Internet as their primary news source (25 percent vs. 19 percent), and Internet users in metroplitan areas did so more frequently than nonmetroplitan dwellers (23 percent vs. 17 percent). Finally, income and education once again revealed significant disparities. While 29 percent of Internet users with family incomes over $100,000 and 31 percent of users with college degrees used the Internet as their primary means of obtaining news, only 16 percent of those with family incomes below $25,000 and 8 percent of those without a high school diploma did so.

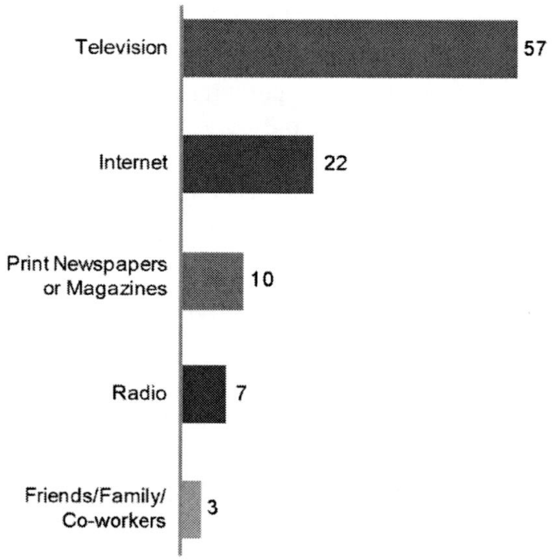

Figure 9. Primary Source of News or Other Information, Percent of Internet Users Age 25+, 2011.

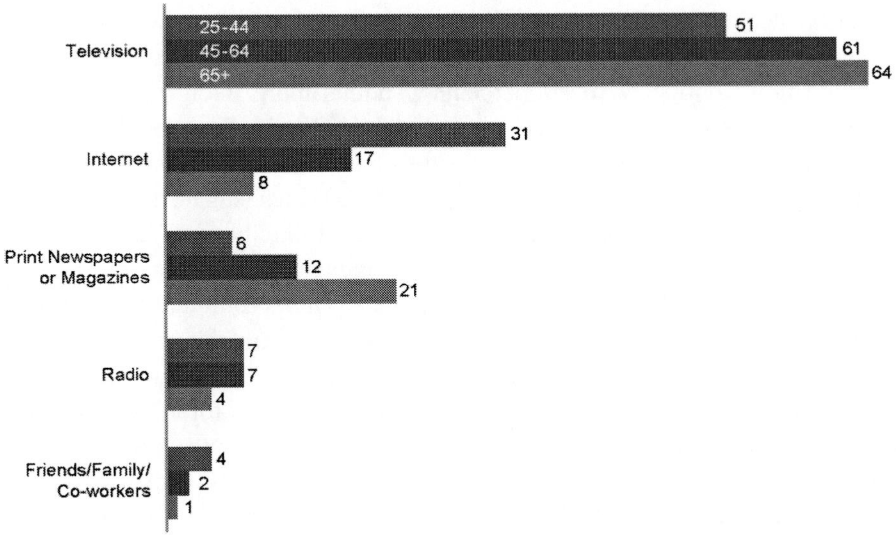

Figure 10. Primary News Source by Age Group, Percent of Internet Users Age 25+, 2011.

The Internet's increasing role in keeping Americans informed of current events prompts examination of whether obtaining news online promotes civic engagement. Several studies suggest a positive association between various dimensions of Internet use and civic engagement (Jennings & Zeitner, 2003; Moy, Manosevitch, Stamm, & Dunsmore, 2005; Tolbert & McNeal, 2003). The CPS Supplement dataset offers one proxy for civic engagement – whether respondents discussed current events with friends, family, or co-workers. Although talking about the news with others was rarely an individual's primary means of obtaining such information, 41 percent of Internet users reported engaging in conversations about the news, compared with just 24 percent of non-users. The CPS data may provide evidence as to whether Internet use, or specifically using the Internet to obtain news, affects the probability that an individual talks about the news with others. Similar to the above analysis of the relationship between Internet use and employment, multivariate regression techniques enable assessing the marginal effects of going online on discussing the news with others when controlling for demographic and other variables. Appendix B presents details of the model (see Table B2).

After accounting for demographic and geographic differences, as well as for use of other news sources, the data demonstrate that Americans who obtained news online were 20 percentage points more likely to discuss the news with friends, family, or co-workers. Interestingly, the model attributes this very powerful correlation to using the Internet for news in particular. In contrast to the earlier analysis of employment, which suggests that merely being an Internet user is associated with increased likelihood of having a job, Internet users did not exhibit increased civic engagement compared with their non-using peers unless they actually obtained news online. Additionally, Americans who engaged in social networking to obtain news or other information were 13 percentage points more likely to discuss the news with others than those who did not using social networking for news. In total, someone who obtained news through social networking sites was 33 percentage points more likely to talk about current events with others than peers who either did not use the Internet or did not utilize it to learn about current events.

The use of traditional media sources for news also appears to correlate with increased propensity to talk about the news with friends, family, or co-workers (see Figure 11). Americans who listened to the radio for news were particularly likely to discuss current events, as they did so at a rate 28 percentage points greater than their non-listening peers. Similarly, those who turned to print media for news were 25 percentage points more likely to talk about the news with others. Watching news on television, however, was associated with a much more modest increase in the probability of discussing current events, as it only boosted the odds by 7 percentage points. These marginal effects are cumulative; Americans who utilized multiple news sources were also more likely to also discuss the news with others. This finding raises the question of how much of the observed effects are due to the various media types themselves, and how much may be attributable to general attentiveness to current events. In other words, individuals who make use of multiple news sources may be more likely to engage in conversations about the news, irrespective of the specific media upon which they depend. Furthermore, utilizing various forms of media appears to have a greater effect on discussing the news with others than demographic factors.[17]

While these findings suggest an association between obtaining news, regardless of source, and talking about the news with friends, family, or co-workers, the results also suggest a positive role for the Internet in encouraging this form of civic engagement. Intuitively, it makes sense that people who talk about the news also stay informed about it through other means, but the model implies that the propensity for someone to engage in such conversations varies based on the news sources they consult. It is encouraging that the Internet may be significantly more effective than television and closer to traditional print media and radio in encouraging civic engagement, particularly in light of the Internet's increasing prominence as a news medium. However, as is true of the earlier regression model examining Internet use and employment, these findings alone cannot establish a causal relationship between online news consumption and civic engagement. The model may not control for important factors influencing civic engagement due to a lack of available data in the CPS.[18] That said, to the extent that obtaining news and other information online might foster increased community participation and social capital, and lead to a more informed electorate, the Internet may strengthen the country's democratic character by encouraging Americans to participate actively in shaping the nation's future.

Mobile Devices

American mobile phone users employ their devices in a variety of ways, as seen in Figure 12. Second in popularity only to making phone calls, 63 percent of users indicated that they sent text messages using their mobile phones. However, with the proliferation of higher quality supplementary features in mobile phones, these devices have become more than a dedicated method of voice or text communication. One-third (33 percent) of mobile phone users browsed the Web with their phones, and the same proportion checked and sent emails using their phones. Furthermore, a little under a quarter of phone owners accessed social networking sites, used GPS and maps, and downloaded applications with their devices (see Figure 12).

Perhaps not surprisingly, younger generations tended to use mobile phones for purposes beyond placing calls. In fact, among mobile phone users under 30 years old, 85 percent used their phones to send text messages and 57 percent browsed the Web with their mobile devices. Those under 30 years old were also the biggest social media users, with 46 percent using their phones to access social networks like Facebook and Twitter compared to 18 percent of those over 45 years old. While some might find high usage by younger people unsurprising, some seniors used their mobile phones for non-call related activities as well. One in four mobile phone users ages 65 and older reported sending text messages, and 9 percent of seniors used their phones to browse the Web.

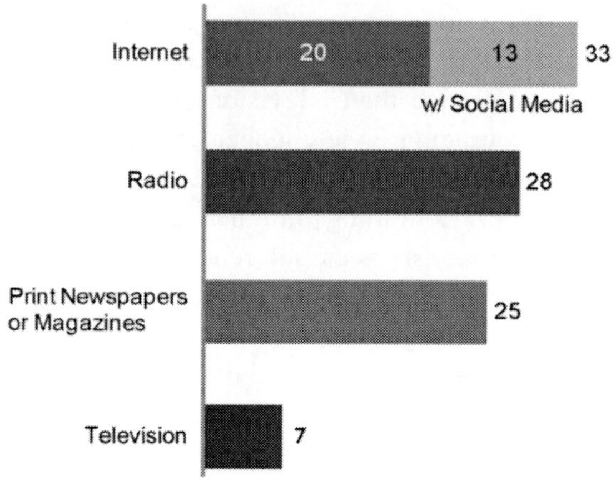

Figure 11. Marginal Effects of News Sources on Discussing News, Percentage Point Increase for Persons Age 25+, 2011.

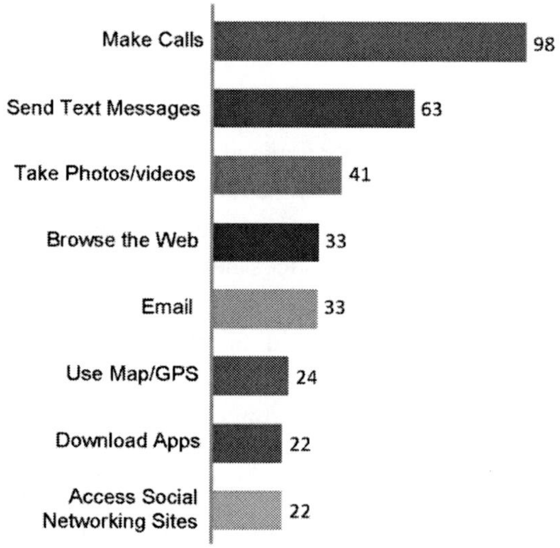

Figure 12. Activities for Which Americans Use Mobile Phones, Percent of Mobile Phone Users Age 25+, 2011.

Slightly more men used a mobile phone than women (87 percent compared to 84 percent, respectively), and men were more likely to use their phones for more specialized activities. Thirty-six percent of men said they used their phone to browse the Web, compared to 31 percent of women. Similarly, 36 percent of men reported using their mobile phones to check email, compared with 31 percent of women. Among mobile phone users, Asian Americans were more likely to use their phones to browse the Web and check email (40 and 41 percent, respectively) than whites (33 percent for each activity) and Hispanics (32 and 31 percent, respectively).

Regular Internet Use

The CPS Supplement results suggest that Americans find great utility in Internet use, as 72 percent of Internet users usually went online every day. An additional 17 percent reported using the Internet at least once per week but not on a daily basis, and the remainder engaged in less frequent use (see Figure 13). The majority of Internet users used the Internet daily irrespective of demographic group, although there were some notable disparities. For example, while 80 percent of Asian American Internet users and 74 percent of their white counterparts went online every day, only 63 percent of African Americans and 64 percent of Hispanics were daily users. Education is an especially prominent source of disparities. While 84 percent of college graduates who went online used the Internet daily, only 72 percent of users with some college credit, 60 percent of those with high school diplomas, and 51 percent of those who did not complete high school reported going online every day. Similarly, 86 percent of Internet users with family incomes of $100,000 or more were daily users, but only 59 percent of those with family incomes below $25,000 used the Internet daily. A significant gap in usage habits existed between Internet users in urban areas and their more rural counterparts, as 73 percent of online Americans living in urban areas went online daily, compared with 66 percent of rural dwellers. When combining daily users with those who go online at least once per week, it is clear that the Internet touches the lives of at least 91 percent of American Internet users on a regular basis. That said, further research into the frequency of Internet use might help determine whether certain factors, such as devices used to go online and locations of Internet use, may affect the benefits people derive from the Internet.

Besides going online frequently, most users also relied on the Internet for certain aspects of their daily lives. In addition to asking a series of questions about various activities Americans undertake using the Internet, the Census Bureau also gathered data on situations in which users consider the Internet an important tool. In other words, while other data show how commonly Internet users go online for activities like job searching and healthcare research, it is also possible to assess to what extent Internet use is essential to performing such functions. Consistent with the Internet's strength as a communications and information tool, large majorities of users stated they relied on it for personal communications (77 percent) and general information (66 percent). Over half of all users relied on the Internet for financial services such as online banking (53 percent), and for shopping and other consumer services (52 percent). Additionally, nearly half of users depended on the Internet for entertainment, and one-third relied on it for on-the-go services such as real-time maps, directions, and destination information (see Figure 14).

The data on frequency of Internet use and reliance on it for a range of activities appear to support the hypothesis that the Internet has become integral to American life. Going online has the potential to enrich both individual lives and entire communities, and increasingly ubiquitous mobile devices have further enhanced the significance of the Internet in Americans' lives. Having considered the benefits of Internet use, the next questions include who connects to the Internet, what motivates use and non-use of the Internet, and how these findings may impact the Administration's goal of universal adoption.

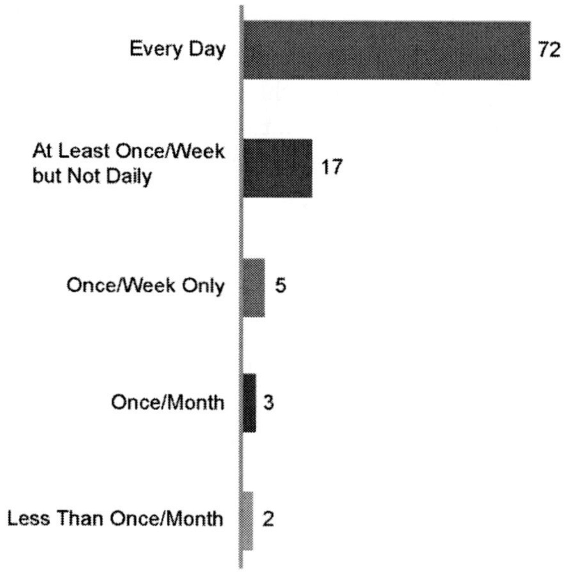

Figure 13. Frequency of Internet Use, Percent of Internet Users Age 25+, 2011.

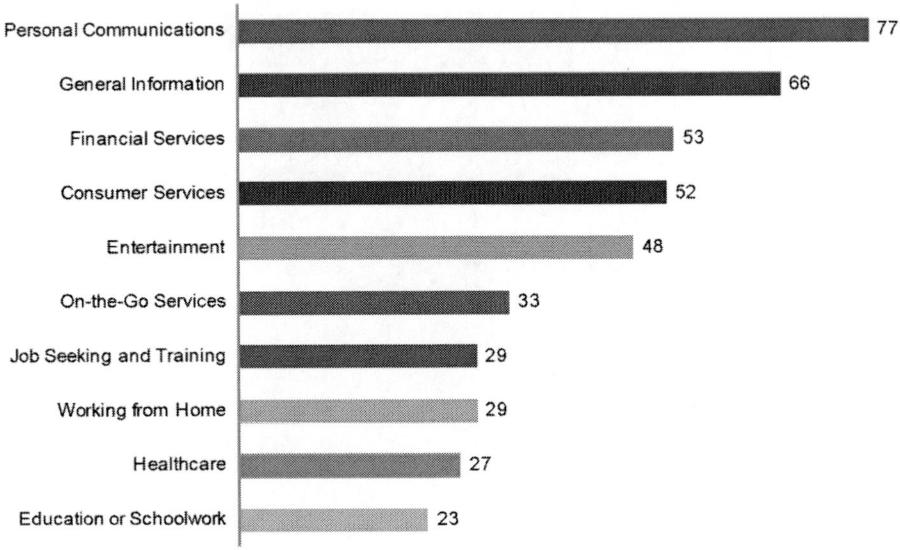

Figure 14. Activities for Which Americans Rely on the Internet, Percent of Internet Users Age 25+, 2011.

3. ADOPTION

Technology has changed in remarkable ways since the U.S. Census Bureau first administered the CPS Computer and Internet Use Supplement in 1994. At that time, policymakers focused on achieving universal telephone subscribership, but the increasing popularity of home computers and modems caused them to begin questioning the impact on communities not connected to the "National Information Infrastructure" ("NII"), as these technologies were then known. Today, ever faster and more advanced means of communication continue to capture the imagination. For example, as noted above and more fully discussed below, a large percentage of households access the Internet using mobile devices, either alone or in combination with some type of personal computer. Moreover, as the number of people using the Internet has increased over time, the technologies used to go online have changed dramatically. While home Internet users relied almost exclusively on dial- up connections during the 1990s, today the vast majority of connected households utilize broadband technologies.

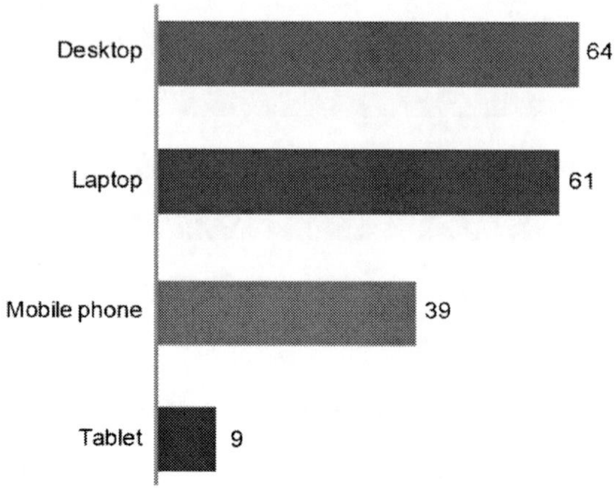

Figure 15. Type of Device Used to Access the Internet, Percent of Internet Users, 2011.

Devices of Choice to Access the Internet

In 2011, the majority of American households owned at least one personal computer. Seventy-six percent of households reported having a desktop, laptop, or tablet computer at home.

As seen in Figure 15, the most common devices people used to access the Internet at any location was a desktop computer (64 percent), followed closely by laptops (61 percent). More than a third of Internet users indicated they accessed the Internet from a mobile phone (39 percent). Tablets were not yet a widespread means of going online for most users in 2011, with only 9 percent saying they accessed the Internet using a tablet.

Table 2 provides more detail on the characteristics of people who used only a personal computer, only a mobile device, or both to access the Internet. People who only accessed the Internet from a smartphone or tablet were younger, on average, than those who only used a personal computer. Compared to Internet users with higher family incomes, those with family incomes below $25,000 were more likely to use only a mobile device to access the Internet. Twenty-nine percent of persons with family incomes below $25,000 used both, compared to 53 percent of those in the $100,000 and above category. A similar pattern held when looking at educational attainment, with less educated users more likely to use only a mobile device. While the incidence of using only a mobile device for Internet access tended to decrease for the higher income and more educated users, and the prevalence of both types of devices tended to increase for higher income and higher educated Internet users, the pattern was different for those Internet users who only used a personal computer. The likelihood of using only a personal computer to go online decreased as family incomes exceeded $50,000, possibly because more educated, wealthier individuals tended to use both their personal computer and a mobile device to go online. While geographic location made little difference for those who used only a mobile device, Internet users in rural locations were more likely to use a personal computer only. The biggest difference by geographic location was for those who used both types of technology to access the Internet – 41 percent of urban dwellers compared to 31 percent of rural dwellers.

Table 2. Personal Characteristics by Type of Device Used to Access the Internet, Percent of Internet Users, 2011

	Mobile Device Only (Mobile Phone, Tablet)	Personal Computer Only (Desktop, Laptop)	Both
All Internet Users	3%	54%	39%
Mean Age (years)	32	39	35
Income < $25,000	6%	57%	29%
Income $25,000-$49,999	4%	59%	32%
Income $50,000-$74,999	2%	56%	38%
Income $75,000-$99,999	2%	53%	43%
Income $100,000 or more	1%	44%	53%
No high school diploma	5%	56%	33%
High school diploma	5%	59%	31%
Some college	3%	50%	44%
College degree or more	1%	45%	52%
Rural	3%	61%	31%
Urban	3%	52%	41%
White	2%	54%	40%
African American	5%	50%	40%
Hispanic	5%	53%	37%
Asian American	2%	53%	43%

Note: Approximately 4 percent of surveyed Internet users did not specify any devices used to access the Internet.

Internet Technologies

The types of technology that households have adopted to access the Internet have changed over time. Dial-up has steadily declined, from 11 percent in 2007 to just 2 percent in 2011, so that virtually all households that use the Internet at home now do so via broadband technologies.[19] A little over a quarter (28 percent) of households do not have Internet access at home, but 69 percent of households have some type of broadband Internet service at home (see Figure 16), compared to only half (51 percent) of households in 2007. Some types of broadband technology – such as DSL and cable modem – are well-established, while more recent innovations include fiber optics and mobile broadband. Of the high-speed Internet technologies, cable modem was the most popular with 33 percent of households, followed by 22 percent of households using DSL.

Income significantly differentiates which households use the Internet at home, as well as the types of technologies they adopted. Just under half (46 percent) of households with incomes of less than $25,000 access the Internet at home. However, the figure rises dramatically for the other income categories. Two-thirds (68 percent) of households earning $25,000-$49,999, 86 percent of households earning $50,000-$74,999, 92 percent of households earning $75,000-$99,999, and 95 percent of the highest income households have Internet access at home.

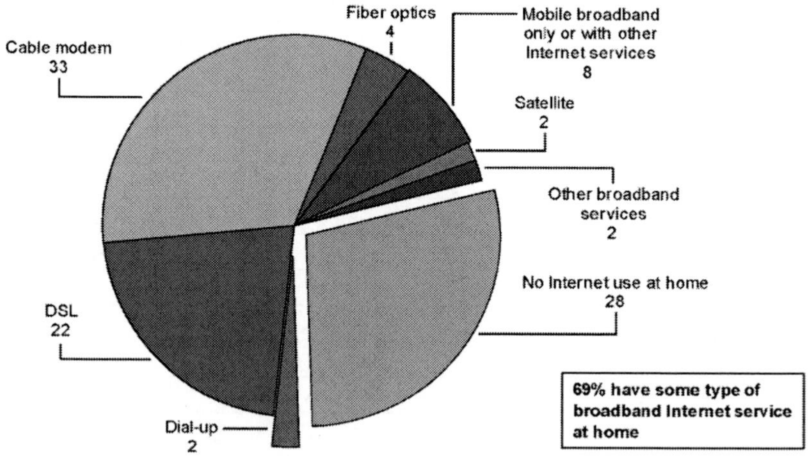

Figure 16. Home Internet Adoption by Type of Technology, Percent of Households, 2011.[20]

Only 14 percent of households with family incomes below $25,000 used DSL, compared to about one-quarter (ranging from 22 to 27 percent) of households in all other income brackets. Mobile broadband, either alone or in combination with other technologies, was not commonly reported regardless of family income, with only 11 percent of households with income $100,000 or greater using it. The most popular technology type for those households in the highest income category was cable modem, with an adoption rate of 46 percent.

NTIA awarded approximately $3.5 billion for **BTOP Comprehensive Community Infrastructure ("CCI")** projects to deploy new or improved broadband Internet facilities (e.g., laying new fiber optic cables or upgrading wireless towers) and connect "community anchor institutions" such as schools, libraries, hospitals, and public safety facilities. As of March 2013, BTOP CCI grantees had deployed about 86,000 miles of new or upgraded network infrastructure. Once completed, these CCI project networks will help to ensure sustainable community growth and provide the foundation for enhanced household and business broadband Internet services. For more information about BTOP, visit NTIA's BroadbandUSA website at http://www2.ntia.doc.gov.

Motivating Factors for Home Internet Service and the Cost of Broadband

Among households using the Internet at home, 37 percent indicated they considered reliable service the most important factor motivating their choice of an Internet service provider (see Figure 17). In other words, households wanted assurance their service would be available to them whenever they needed it. Connection speed closely followed reliability, with one-third (33 percent) of households identifying speed as the most important factor. Almost one-quarter (24 percent) of households deemed affordability the most important factor in selecting their providers.

The types of Internet technologies that households adopt provide different benefits and shortcomings. Figure 18 depicts the breakdown of which factors households deem most important by technology type. Affordability is a much more important factor for dial- up users than for broadband users – 48 percent of dial-up users indicated they were most concerned about the cost of Internet service, compared to 24 percent overall. Similarly, and not surprisingly, dial-up users valued connection speed less than Internet-using households at large – 15 percent as compared to 33 percent of all such households.

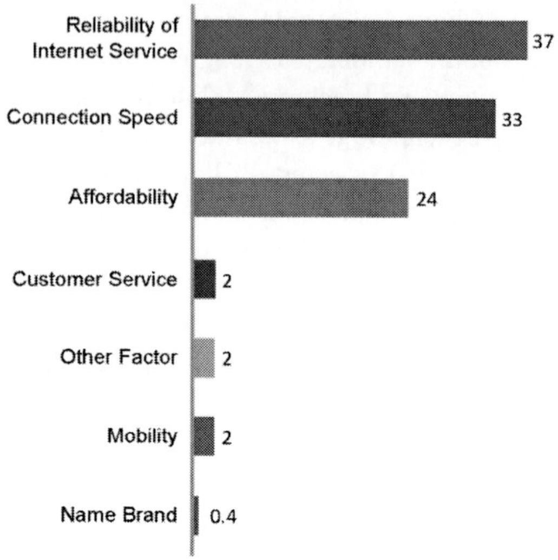

Figure 17. Most Important Factors for Home Internet Service, Percent of Online Households, 2011.

Among satellite users, connection speed was the most important factor. Of the various Internet connection technologies, satellite is not especially fast, but for communities that lack a wired broadband option (such as DSL or cable), satellite is much faster than the alternative dial-up technology. As depicted in Figure 18, more than a third (39 percent) of satellite users considered connection speed the most important factor, the highest percentage of any technology type to indicate connection speed as most important, and 6 percentage points higher than connected households as a whole. Cable modem users also valued connection speed more than the average for online households, with 35 percent indicating it as the most important factor. Predictably, mobility was an important factor for users of mobile broadband technology, with 9 percent of mobile broadband users considering it the most important factor, compared to only 2 percent of all Internet-adopting households. While that might seem low considering the untethered nature of mobile broadband is a fundamental characteristic of the technology, it is reasonable to assume that mobility is more of an expected factor than the most important. At 34 percent, a plurality of mobile broadband users indicated they thought reliability of Internet service to be the most important factor. Forty-two percent of fiber optics users thought reliability was the most important factor, compared to 37 percent of all Internet adopters as presented in Figure 17.

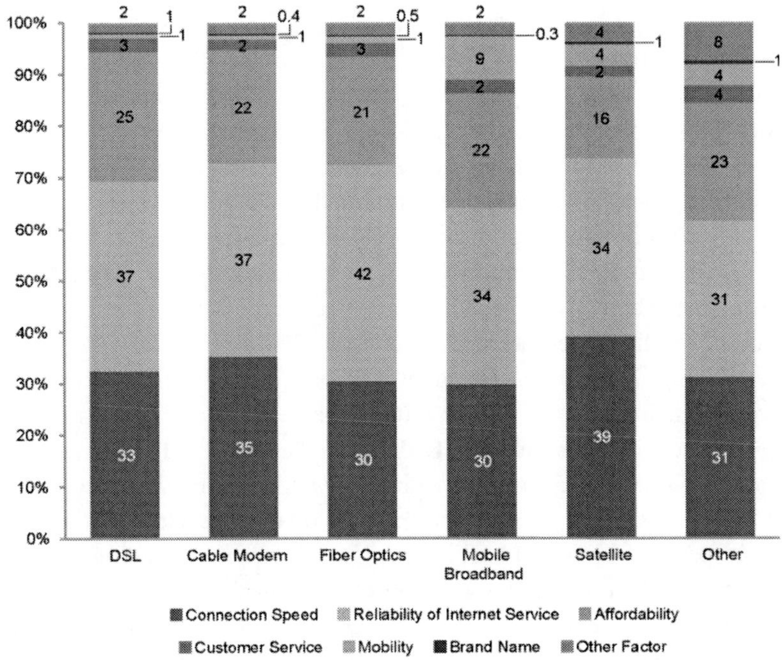

Figure 18. Most Important Factors for Home Internet Service by Technology Type, Percent of Online Households, 2011.

In areas served by multiple Internet service providers, households have the ability to switch if they are not satisfied with their current service. Figure 19 depicts the main reasons why households that switched Internet service providers did so. Thirty- eight percent of those households specified that they switched to less expensive services, followed by 30 percent of households that changed because their new provider offered faster connection speeds. Although reliability was the most important factor for households when choosing a provider, only one in ten households (10 percent) indicated that they had switched providers for reliability reasons.

Since affordability was the most-cited factor for households switching providers, how much households said they pay for their Internet service is a key question. Figure 20 lists the cost of high-speed Internet per month by technology type for those households subscribing to Internet service as a standalone (non-bundled) product. Over a third (36 percent) of households with broadband Internet reported they paid between $30 and $44 per month. Only 6 percent of households with broadband reported paying less than $15 per month, while 16 percent reported that they paid $60 or more per month.

Satellite users appear to have spent the most for their Internet per month, with 31 percent reporting paying $60 or greater. On the other hand, DSL using households were least likely to pay $60 or more for Internet service, and 70 percent of DSL users paid less than $45 per month.

Two-thirds (66 percent) of households with broadband Internet service pay for their subscription through a bundled package. Fifty-eight percent of households with broadband Internet reported paying between $25 and $49 per month for the Internet portion of their service bundle (see Figure 21). Only 9 percent of households with broadband responded that they paid less than $25 per month for the Internet portion of their bundle, while just 11 percent reported paying $75 or more per month. Fiber optics users spent slightly more money on their Internet per month, with 13 percent paying $75 or more for the Internet portion of their bundle. DSL users had the smallest percentage of households that paid $75 or more with just 9 percent, and 72 percent paid less than $50 per month for the Internet portion of their service bundle. Because the prices paid by households for broadband Internet service and for service bundles are self-reported, the data may be subject to reporting error. The results may also be influenced by unknown behavioral factors, such as subscribers to bundled services preferring faster speeds or being upsold to more expensive plans.

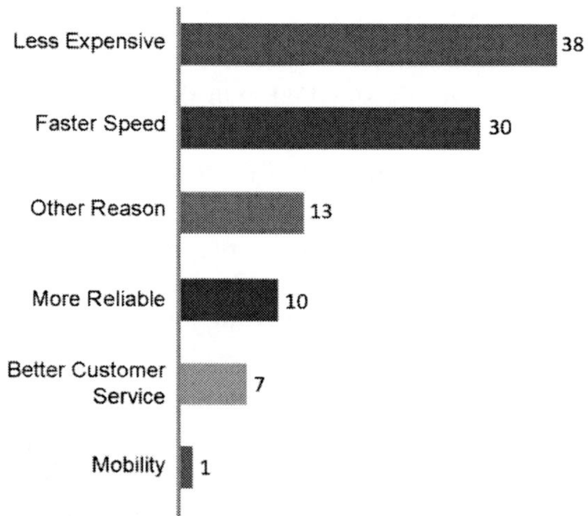

Figure 19. Main Reason for Switching Internet Service Providers, Percent of Switching Households, 2011.

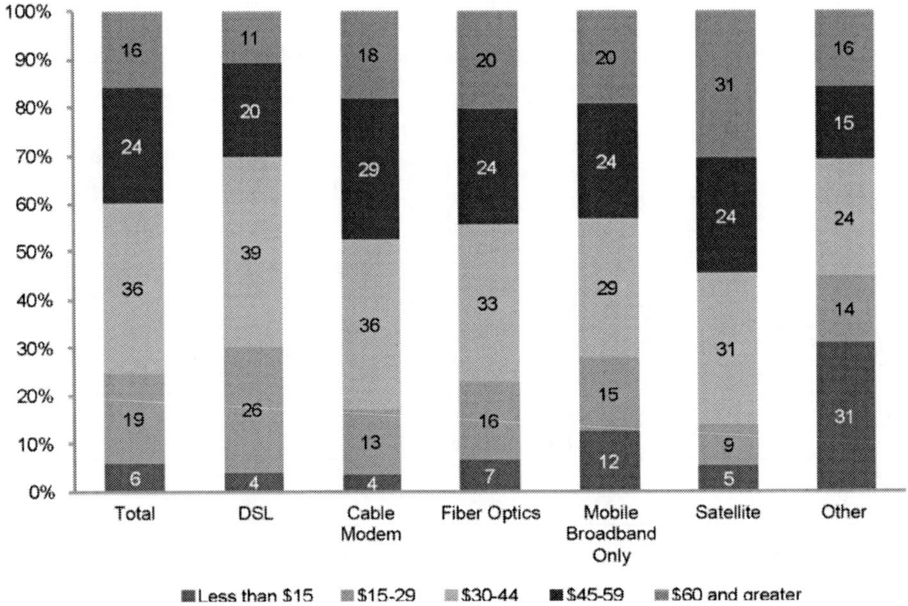

Figure 20. Cost of Broadband Internet per Month by Technology Type, Percent of Online Households Subscribing to Non-Bundled Internet Service, 2011.

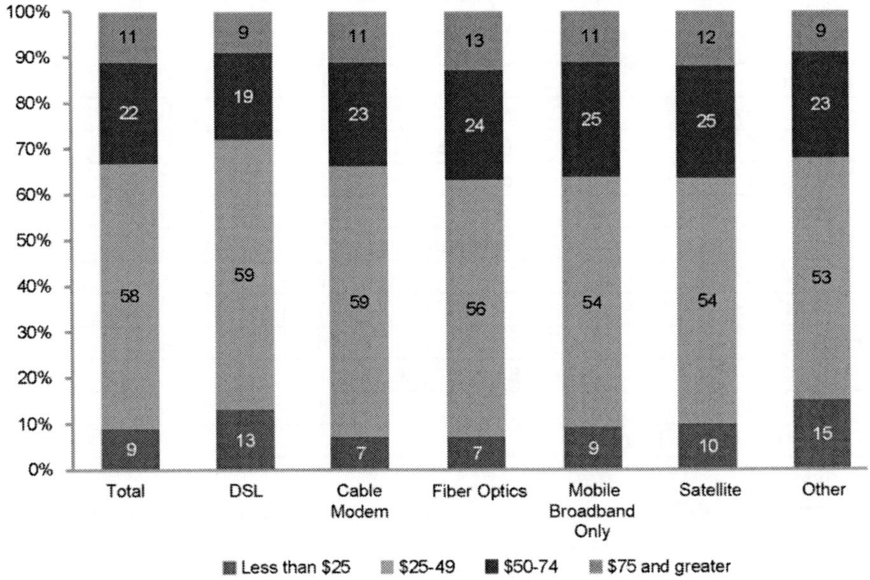

Figure 21. Cost of Broadband Internet Portion of "Bundle" per Month, Percent of Online Households Subscribing to Bundled Internet Service, 2011.

Table 3. Home Computer, Internet, and Broadband Adoption by Demographic Characteristics and Geographic Location, Percent of Households, 2011

	Computer Ownership	Internet Use	Broadband Adoption
All households	76%	72%	69%
Family Income			
Income < $25,000	52%	46%	43%
Income $25,000-$49,999	73%	68%	65%
Income $50,000-$74,999	89%	86%	84%
Income $75,000-$99,999	93%	92%	90%
Income $100,000 or more	95%	95%	93%
Householder Education			
No high school diploma	43%	37%	35%
High school diploma	65%	61%	58%
Some college	82%	77%	75%
College degree or more	92%	90%	88%
Metropolitan Status			
Rural	67%	62%	58%
Urban	77%	74%	72%
Householder Race and Ethnicity			
White	80%	76%	74%
African American	62%	57%	55%
Hispanic	63%	58%	56%
Asian American	85%	83%	81%
Household Type			
Households with school-age children	84%	81%	79%
Households without school-age children	73%	69%	66%
Householder Age			
16 to 44 years	82%	79%	77%
45 to 64 years	79%	76%	73%
65 years and older	56%	52%	49%
Householder Disability Status			
Has a disability	53%	48%	46%
Does not have a disability	79%	76%	73%

Computer and Broadband Internet Adoption by Demographic Characteristics and Geography

Demographic and geographic characteristics of U.S. households likely play a large role in determining a household's likelihood of owning at least one computer. These characteristics also may shed light on overall computer ownership and broadband adoption rates for U.S. households.

Table 3 demonstrates how demographic characteristics and geographic location can influence the propensity for computer ownership, Internet use, and broadband adoption for households. Low- income and less educated households have well below average computer ownership and broadband rates. Less than half (43 percent) and only about a third (35 percent) of households headed by someone with less than a high school diploma have a computer at home or broadband access, respectively. Furthermore, fewer non-Asian minority households adopted broadband Internet in 2011 (55 percent of African American and 56 percent of Hispanic households), compared to all households in the United States (69 percent) and to white and Asian American households (74 percent and 81 percent, respectively). Households headed by someone with a disability also had lower levels of computer and Internet use.

Households with school-age children were more likely than households without children in school to own a computer (84 percent, compared to 73 percent) and adopt broadband (79 percent, compared to 66 percent). Households headed by someone under the age of 45 had higher rates of home computer ownership (82 percent) and broadband adoption (77 percent), compared to their older counterparts.

Among all households, those located in an urban area were more likely to own a computer and adopt high-speed Internet than rural households. As indicated previously, 76 percent of all U.S. households owned at least one computer, and 69 percent had some form of broadband technology. Urban households had higher rates of computer ownership (77 percent) compared to their rural counterparts (67 percent). As for broadband use, urban households similarly had higher adoption rates (72 percent) compared to rural households (58 percent).

Compared to the national figure, white and Asian American households had higher computer ownership rates in both urban and rural areas, while African Americans, Hispanics, and American Indian and Alaska Native ("AIAN") households had lower ownership rates in both areas. In addition, for African American and AIAN households, there was a pronounced gap between urban and rural computer ownership, with a difference in computer ownership rates for urban and rural African American households of 16 percentage points, and 25 percentage points for AIAN households (see Table 4). Less than two-fifths of rural Hispanic, African American, and AIAN households reported having broadband Internet, and just 57 percent of urban African American households had high- speed Internet in 2011. While only about half of urban and rural households earning less than $25,000 owned a computer (53 percent and 47 percent, respectively), computer ownership was only slightly less than the national rate for households with family incomes between $25,000 and $50,000. By contrast, households with incomes of $50,000 and greater reported very high rates of computer ownership. Compared to the national broadband adoption figure, rural broadband use for households with incomes less than $25,000 was particularly low (36 percent) compared to their counterparts with incomes of at least $100,000 (86 percent). However, both urban and rural households with incomes above $50,000 adopted broadband at rates higher than the national adoption figure for each location. Lastly, there was a positive relationship between educational attainment and computer ownership and broadband adoption for both urban and rural households. Broadband adoption was quite low for rural households headed by someone without a high school diploma (27 percent), but grew with educational attainment to 80 percent among householders with a college degree or more.

Table 4. Household Computer Ownership and Broadband Internet Adoption by Urban/Rural Location, Race, Ethnicity, Income, and Education, Percent of Households, 2011

Household Characteristic	Computer Ownership		Broadband Adoption	
	Urban	Rural	Urban	Rural
All Households	77%	67%	72%	58%
Race and Ethnicity				
White	82%	71%	77%	61%
African American	64%	48%	57%	35%
Hispanic	64%	51%	58%	39%
Asian American	85%	86%	81%	81%
American Indian and Alaska Native	69%	44%	60%	33%
Family Income				
Less than $25,000	53%	47%	45%	36%
$25,000-$49,999	74%	69%	67%	59%
$50,000-$74,999	90%	85%	85%	77%
$75,000-$99,999	94%	91%	91%	83%
$100,000 or more	96%	92%	94%	86%
Education				
No high school diploma	45%	36%	37%	27%
High school diploma	66%	61%	60%	51%
Some college	82%	78%	77%	68%
College degree or more	93%	88%	89%	80%

Table 5. Household Computer Ownership and Broadband Internet Adoption by Urban/Rural Location, Race, Ethnicity, and Income, 2011

Household Characteristic	Computer Ownership		Broadband Adoption	
	Urban	Rural	Urban	Rural
All Households	77%	67%	72%	58%
White				
Less than $25,000	59%	50%	51%	38%
$25,000-$49,999	77%	71%	70%	61%
$50,000-$74,999	91%	86%	86%	79%
$75,000 or more	96%	92%	94%	85%
African American				
Less than $25,000	45%	37%	37%	26%
$25,000-$49,999	67%	63%	59%	45%
$50,000-$74,999	87%	74%	82%	58%
$75,000 or more	88%	69%	86%	69%
Hispanic				
Less than $25,000	44%	40%	36%	31%
$25,000-$49,999	66%	51%	58%	36%
$50,000-$74,999	84%	81%	81%	66%
$75,000 or more	89%	71%	86%	60%

Table 5 depicts the relationship of race and ethnicity, income, and geographic location to the propensity for computer ownership and broadband adoption in households in 2011. Rural African American and Hispanic households with annual family incomes below $25,000 reflected the lowest computer ownership rates (37 percent and 40 percent, respectively) and broadband adoption rates (26 percent and 31 percent, respectively). There was a large geographic gap for white households with incomes below $25,000 – just over half (51 percent) of white urban households reported broadband adoption, compared to just 38 percent of white rural households.

Table 6 contains data on computer use and broadband adoption by race and ethnicity, educational attainment, and geographic location. Similar patterns emerge when examining educational attainment and family income. In 2011, less than a quarter (24 percent) of African American rural households headed by someone without a high school diploma said that they owned a computer, compared to 67 percent of all rural households, and 76 percent of all U.S. households. Concerning broadband service, only 16 percent of African American rural households headed by someone without a high school diploma used high-speed Internet at home. In general, for each geographic and demographic category, the difference in computer ownership rates and broadband adoption rates for householders without a high school diploma compared to those with a high school diploma was between 10 and 20 percentage points.

As the tables above show, geographic location matters considerably to broadband adoption rates. Figure 22 demonstrates how the types of Internet technology vary for urban and rural households. The urban-rural gap in Internet use at home was 11 percentage points in 2011, with 27 percent of urban households and 38 percent of rural households indicating they did not use the Internet at home (compared to 28 percent for all U.S. households). While dial-up use was low regardless of location, cable modems were much more common in urban households (35 percent), and DSL and satellite were slightly more common in rural households (26 percent and 3 percent, respectively).

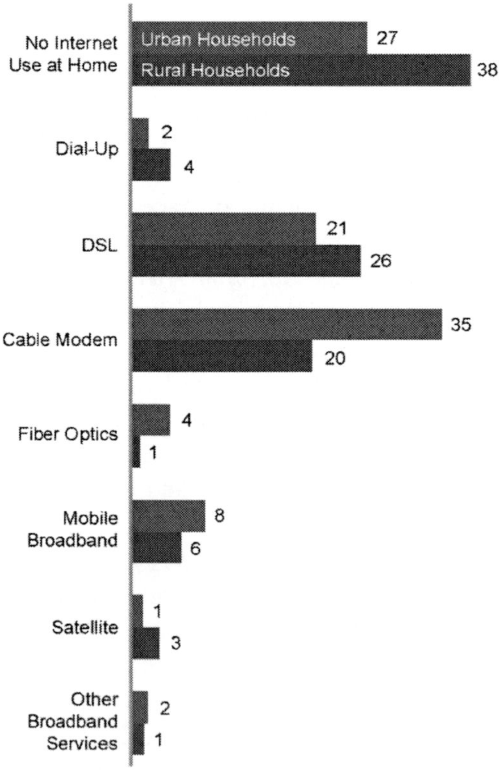

Figure 22. Internet Connection Type by Urban and Rural Location, 2011.

Table 6. Household Computer Ownership and Broadband Internet Adoption by Urban/Rural Location, Race, Ethnicity, and Education, 2011

Household Characteristic	Computer Ownership		Broadband Adoption	
	Urban	Rural	Urban	Rural
All Households	77%	67%	72%	58%
White				
No high school diploma	49%	39%	41%	29%
High school diploma	70%	64%	64%	54%
Some college	85%	79%	80%	70%
College degree or more	94%	89%	90%	81%
African American				
No high school diploma	34%	24%	28%	16%
High school diploma	53%	41%	46%	29%
Some college	71%	69%	64%	54%
College degree or more	88%	87%	83%	63%
Hispanic				
No high school diploma	46%	37%	36%	26%
High school diploma	61%	48%	56%	36%
Some college	80%	76%	73%	60%
College degree or more	86%	84%	81%	78%

There was also considerable variation in computer ownership and Internet use among states (see Figure 23). The proportions of households with a computer at home ranged from 65 percent in Mississippi to 85 percent in Washington. Average home broadband adoption ranged from 53 percent in Mississippi to 80 percent in Washington.[21]

Within state boundaries, policymakers have focused considerable attention on relative rates of broadband use in rural versus urban areas. As Table 3 demonstrated, broadband adoption rates tend to be higher in urban locations than rural locations. Figure 24 depicts broadband adoption rates in urban and rural locations by state. The size of the urban-rural gap ranged from just 1 percentage point in Delaware, to 29 percentage points in Virginia. Alaska had the lowest estimated rural broadband adoption rate at 41 percent, while New Hampshire had the highest estimated rural broadband adoption rate at 75 percent.[22]

Household Computer Ownership by State, Rounded to Nearest Percentage Point, 2011

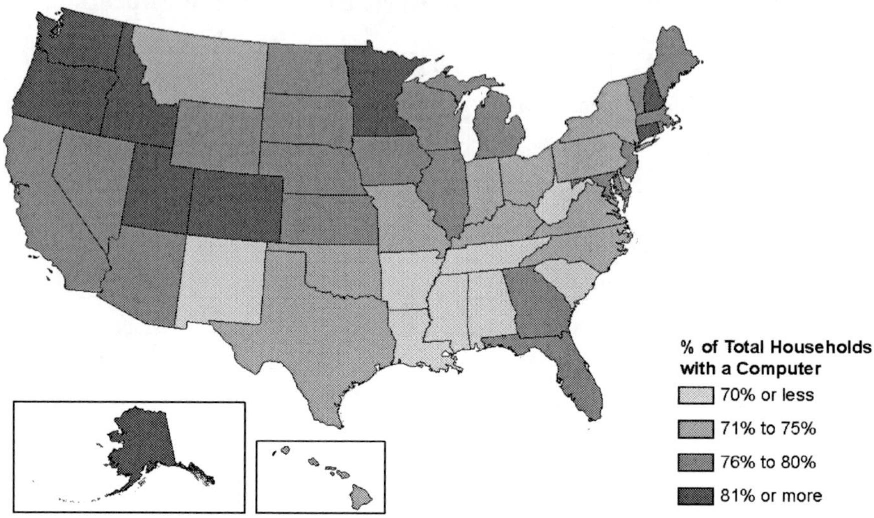

Household Broadband Internet Adoption by State, Rounded to Nearest Percentage Point, 2011

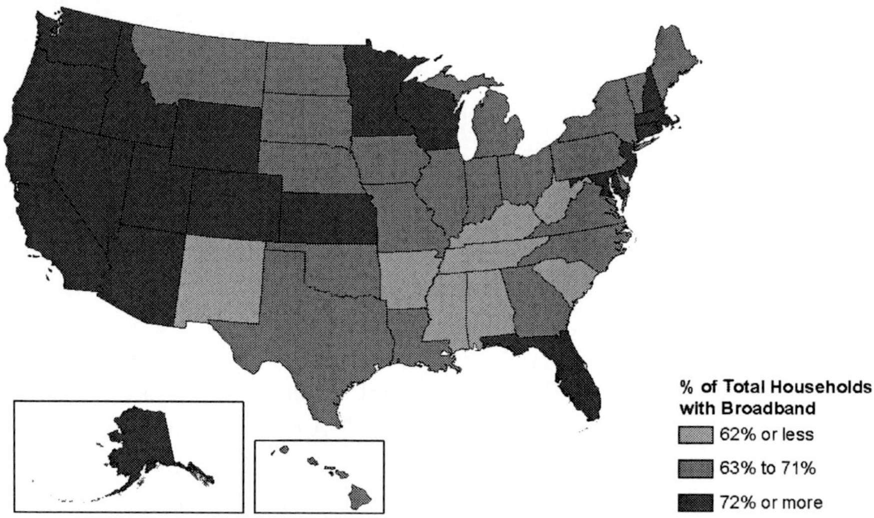

Figure 23. Computer and Internet Use by State (by Household Broadband Adoption Rate).

Broadband Internet Use in Urban Locations by State, Rounded to Nearest Percentage Point, 2011

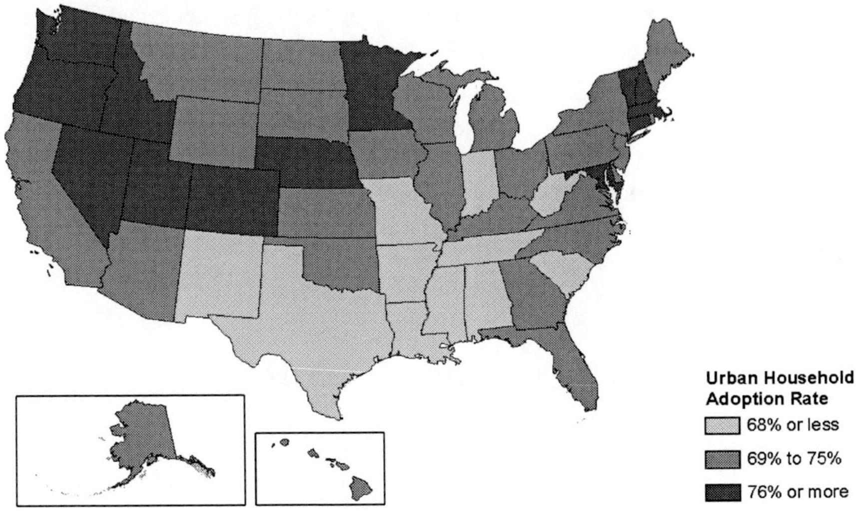

Broadband Internet Use in Rural Locations by State, Rounded to Nearest Percentage Point, 2011

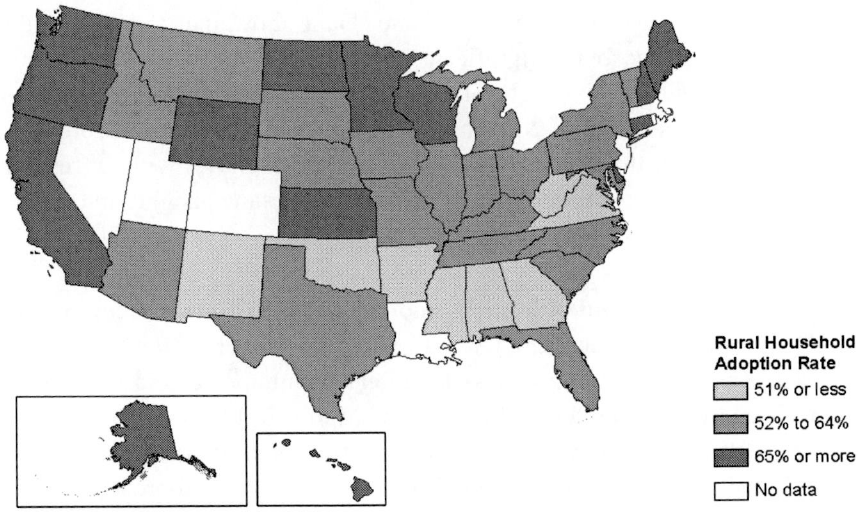

Figure 24. Broadband Internet Use in Urban and Rural Locations (by Overall Household Adoption Rate).

Adoption Outside the Home

Nine out of ten people (92 percent) who said they accessed the Internet used it from home. However, those Internet users who do not have a computer at home may go online at another person's house, work, public libraries, and other places that offer an Internet connection. As seen in Figure 25, about a third of such users (32 percent) reported using the Internet at a public library. Internet users without a computer at home were also likely to go online at someone else's house (29 percent), school (25 percent), or their place of work (23 percent).

4. NON-ADOPTION AT HOME

Despite the declining cost of computers, the proliferation of mobile devices and services, and the availability of wired broadband Internet service to 93 percent of Americans (NTIA & FCC, 2013),[23] many individuals remain disconnected from the digital world. Whether by choice or by circumstance, in 2011 about 30 percent of the 119 million American households represented in the CPS did not use broadband at home. Data show that the digital divide continues to narrow, but a significant gap remains. As in past years, survey respondents identified a variety of reasons for not connecting to the Internet where they live (see Figure 26). These reasons – discussed in rank order based on responses from the 2011 survey's total U.S. households without residential broadband or dial-up service – may provide important insights into effective strategies for promoting universal connectivity.

NTIA Broadband Adoption Toolkit: BTOP grantees received about $452 million of Recovery Act funding to develop new broadband subscribers by expanding access to public computer centers and training individuals to use the technology. As of March 2013, BTOP investments have yielded 41,000 new workstation installations, more than a half million new broadband subscribers, and 4.2 million newly trained broadband users who received more than 12.3 million hours of training. Based on the extensive experience of BTOP projects in about 100 communities across the nation, NTIA released in May 2013 a toolkit of practical ideas and resources grantees have used successfully to help overcome barriers that prevent Americans from adopting broadband at home. The toolkit is available at http://www2.ntia.doc.gov/files/toolkit_042913.pdf.

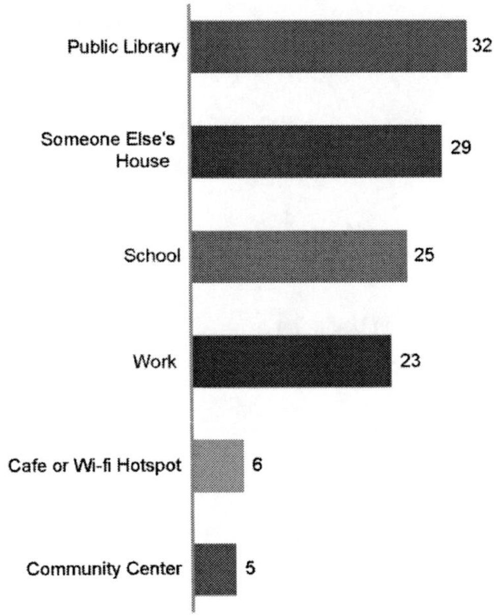

Figure 25. Internet Use Locations Outside the Home, Percent of Internet Users Without a Computer at Home, 2011.

No Need or Interest

In 2011, 48 percent of all households that did not use the Internet at home reported that they primarily had no need for or interest in going online. Fifty-two percent of households that have never used the Internet at home stated they lacked a reason to do so, while 34 percent of dial- up users stated they did not need residential high-speed Internet service, though presumably the latter group found sufficient utility in accessing the Internet at slower speeds. Approximately 20 percent of former home Internet users cited a lack of need or interest as the most important reason for discontinuing their use.

With respect to income, households lacking an Internet connection with annual family incomes of at least $100,000 were the most likely (at 53 percent) to express disinterest in using the Internet at home, compared to only 45 percent of those households with incomes less than $25,000. In addition, 8 percent of those highest earning households without Internet at home reported that they could go online elsewhere. By contrast, only 2 percent of non-adopting households reporting annual incomes below $25,000 said they could use the Internet at locations other than home.

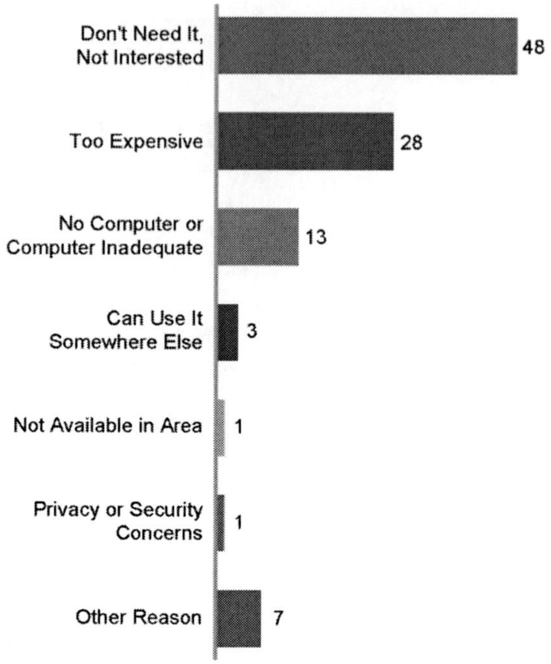

Figure 26. Main Reason for Non-Use of the Internet at Home, Percent of Households Not Online at Home, 2011.

Householder age also pointed to strong differences in perceptions about the necessity of home Internet service. Sixty- seven percent of users 65 years and older cited a lack of need as the reason for not using the Internet at home, compared to 26 percent of those between ages 16 and 44 and 46 percent of those between the ages of 45 and 64. Disability status was also a factor. Fifty-one percent of householders with a disability who did not use the Internet at home stated they had no need to use it, which represented a 3-percentage point decline from 54 percent in 2010. While reasons for the decrease are not clear, the result may point to a developing trend of increased home Internet use by persons with disabilities.

White households (53 percent) and Asian American households (56 percent) replied most often that they did not need to go online at home. Non-Asian minority households, however, were less likely to cite a lack of need as a reason for not using home Internet services, with only 40 percent of African American households and 39 percent of Hispanic households responding they did not need or want to use the Internet at home (see Figure 27).

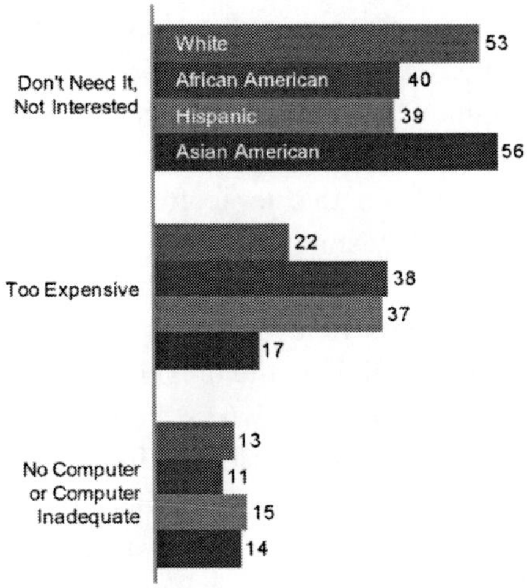

Figure 27. Main Reason for Non-Use of the Internet at Home by Race, Percent of Households Not Online at Home, 2011.

The presence of school-age children appears to have affected households' views about the importance of adopting the Internet in the home. Families without children living at home were more than twice as likely as those with children to cite lack of interest as the primary reason they did not have residential Internet service (52 percent compared to 26 percent). This gap narrowed significantly for dial-up households explaining why they did not use broadband at home, with 36 percent of those without children citing no need or interest for home high speed Internet service, while only 23 percent of dial-up households with children answered similarly.

Communication Service for the Deaf, Inc. ("CSD") received a $15 million BTOP grant to increase broadband adoption by deaf and hard of hearing people. CSD's Project Endeavor seeks to cultivate an estimated 16,000 new broadband subscribers throughout the U.S. and its territories by offering discounted broadband service, specialized computers, and technology training. In addition, CSD is providing online employment courses, sign language interpretation, captioned video services, and other content and functionalities designed to advance the educational, employment, and healthcare interests of project participants. More information on this grant can be found at http://www.ntia.doc.gov/press-release/2010/commerce-departments-ntia-announces-recovery-act-investment-expand-broadband-inte.

These data may aid researchers and policymakers in assessing whether children's use of the Internet in schools and libraries has resulted in increased home broadband adoption, as children's exposure helps demonstrate its relevance to family members unfamiliar with the technology. Such findings are consistent with research on this topic from other countries, including a recent study in Portugal suggesting that "school broadband use contributes directly to a higher adoption rate in households with children" (Belo & Ferreira, 2012). Studies in this field may bolster justification for the FCC's goal of "upgrading and modernizing" the universal service support mechanism for schools and libraries, popularly known as the E-rate program (FCC, 2010).

Householder employment status also significantly impacted whether the household lacked residential Internet service because of a belief the service was unnecessary. By a margin of 35 percentage points, non-adopting households in which the head was not in the labor force, such as retirees, homemakers, and discouraged job seekers who have ceased searching for work were most likely to cite no need or desire to use the Internet (57 percent), compared to 22 percent of households led by an unemployed householder. Thirty nine percent of employed householders without Internet at home stated they had no need for home online connectivity, and they were twice as likely as their unemployed counterparts to have cited the ability to use the Internet elsewhere (6 percent compared to 3 percent).

In addition, place of household residence affected opinions about the need or desire to access the Internet at home. Fifty percent of rural households without the Internet at home expressed that they had no need for, or interest in, such use compared to 47 percent of urban households that held that view.

Expense

In the nine months between the 2010 and 2011 CPS data collections, the percentage of households responding that the expense of Internet service prevented them from using it at home increased to 28 percent in 2011 from 24 percent in 2010. Further, 37 percent of dial-up customers cited the high cost of broadband to explain why they had no such service at home. Former home Internet customers were the most price-conscious. By a margin of two to one, previous users cited expense as the predominant reason they discontinued service at 41 percent, compared to 20 percent who no longer accessed the Internet at home because they did not need the service. In reverse order, but by a similar margin, 26 percent of households that had never accessed the Internet at home identified cost as their main concern, while 52 percent of such households reported they did not need the service (see Figure 28).

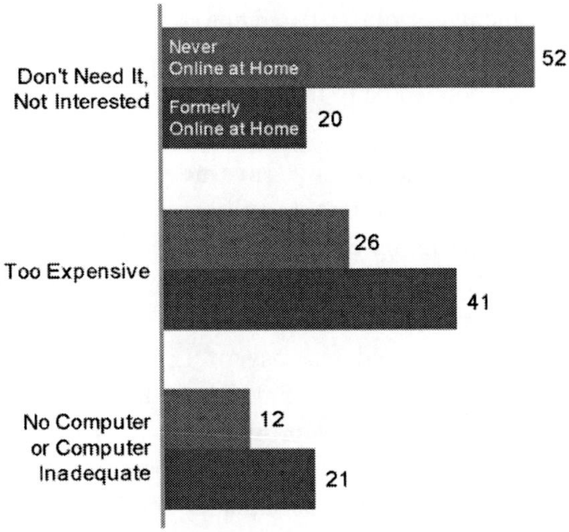

Figure 28. Main Reason for Non-Use of the Internet at Home by Previous Home Use, Percent of Households Not Online at Home, 2011.

Not surprisingly, family income levels are also associated with different reasons for non-use of the Internet at home. Thirty-two percent of households with annual income below $25,000 cited the high cost of Internet service, compared to 18 percent of those of householders with annual incomes above $50,000. For households that once, but no longer, used the Internet at home because of the expense, the figure soared to 51 percent of those with annual incomes less than $25,000.

Each year, the Census Bureau develops poverty thresholds based on family size and economic conditions in the calendar year to calculate the income at which a family would be classified as living in poverty that year. In 2011, the Census Bureau established weighted average poverty thresholds ranging from $11,484 to $23,021 for families of one to four members (U.S. Census Bureau, 2012). Using these benchmarks, but without regard to family size, as general approximations of poverty thresholds, almost 52 percent of all households without any home Internet at all or with only dial-up service had annual incomes less than $25,000 in the July 2011 CPS survey. While federal poverty guidelines are complex and vary by program – making it infeasible to identify those households in the CPS sample that were living in poverty – the CPS data do reveal that 62 percent of households responding that the affordability of Internet service was the primary deterrent to their home Internet use also reported incomes below $25,000. Moreover, among households in that income group, 51 percent of former users ceased using the Internet at home because the service became too expensive, and 30 percent of those who have never connected to the Internet in their homes cited high costs.

The Lifeline Program for Low-Income Consumers offers discounts on landline or pre-paid wireless telephone service to consumers with income at or below 135 percent of the federal Poverty Guidelines or enrolled in certain federal assistance programs. As part of its effort to modernize this universal service program, the FCC has established a $14 million pilot program to field test use of Lifeline subsidies to increase broadband adoption among nearly 75,000 low- income consumers without such service. The 18-month pilot program consists of 14 projects in 21 states and Puerto Rico that must provide participating consumers with subsidized standalone or bundled broadband service for 12 months using wired or wireless technologies.

The importance of expense as a primary reason for not using the Internet at home diminished as the householder's age increased. Households of those between ages 16 and 44 were most likely to cite this reason at 47 percent of non-users, while only 10 percent of 65 and older such households did so.

The proportion of African American and Hispanic households that stated they did not use the Internet in their homes because it was too expensive registered at 38 percent and 37 percent of non-adopters, respectively. These two groups experienced the largest increase in this response from 2010, when 30 percent of African American non-using households and 35 percent of their Hispanic counterparts reported cost as the most important reason for not accessing the Internet at home. Twenty-two percent of white non-adopting households explained cost as their main reason for no residential Internet service. Asian American households lacking home Internet service were least likely at 17 percent to say that cost kept them from accessing the Internet at home.

The presence of school-age children living in a household appeared to be a dividing factor for reasons why households did not use residential Internet. Only 24 percent of non-adopting households without school-age children determined that home Internet access was not affordable compared to 49 percent of such households with children under 18 years old.

A significant gap existed between non-using households with employed (33 percent) and unemployed (55 percent) householders who explained their lack of home Internet service as primarily a matter of cost. By contrast, only

21 percent of households lacking home Internet in which the householder was out of the labor force responded that the expense of residential Internet service was the main reason for not using it (see Figure 29). While 29 percent of non-using urban households identified price as their main reason for no residential Internet service, only 25 percent of their rural counterparts did so.

Lack Adequate or Any Computer

Approximately 13 percent of households without home Internet services cited their lack of an adequate computer or any computer at all as their main reason for not accessing the Internet at home. This represents a 2-percentage point decline from 2010, when 15 percent of such households offered this explanation. Income did not appear to be a significant factor influencing households' responses about the lack of a computer as the main reason for no home Internet use, with only 3 percentage points separating non-adopting households earning less than $25,000 at 13 percent from those with incomes of $100,000 or more at 10 percent. It is interesting to note, however, that of all the unconnected households without a computer or a well-functioning one, 57 percent earned less than $25,000 annually and 30 percent earned between $25,000 and $50,000, while only 13 percent of such households citing this reason reported incomes exceeding $50,000 per year.

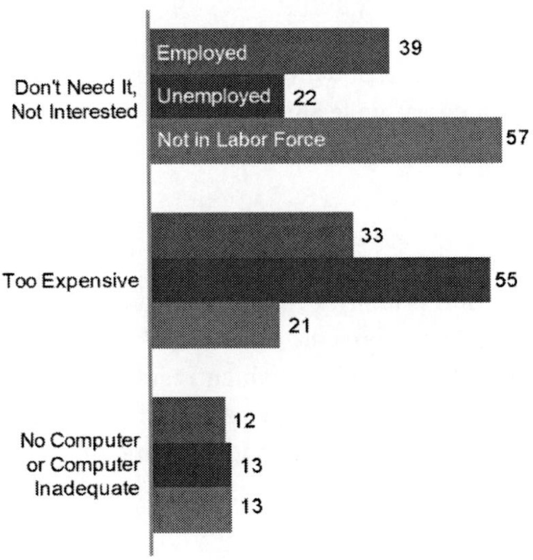

Figure 29. Main Reason for Non-Use of the Internet at Home by Employment Status, Percent of Households Not Online at Home, 2011.

Hispanic households with home Internet service identified the lack a computer or an inadequate one as the primary reason for not using the Internet at home most often at 15 percent, while their African American counterparts did so least often at 11 percent. Fifteen percent of non-using households headed by a person with a disability stated they did not own a computer or one that worked well enough to access the Internet at home compared to 12 percent of those whose head of household was not disabled.

Connect 2 Compete ("C2C") is an initiative of the cable industry, national retailers, non-profit organizations, and others to close the digital divide. Families with children participating in the National School Lunch Program may subscribe to high-speed Internet service for $9.95 per month and purchase for $150 a refurbished laptop or desktop computer, including software and a 90-day warranty and technical support. They may also obtain in-person or online digital literacy training.

Interestingly, 21 percent of former residential Internet users cited not having a suitable computer as the primary reason for not accessing the Internet in their homes. In addition, examining former home Internet users by racial and ethnic category shows that 32 percent of Asian American and 26 percent of Hispanic households in this group responded that not having an adequate computer was their main reason why they no longer accessed the Internet at home. Yet, only 20 percent of white and 21 percent of African American households in this category gave this reason to explain why they no longer went online in their homes.

Other Reasons

In 2011, 7 percent of householders without home Internet service stated they lacked a connection primarily for "other reasons." Although respondents reported a range of other reasons, the data do not reveal discernible patterns in the responses. Among households in which residents formerly went online at home, 13 percent stopped for "other reasons." By comparison, 6 percent of households that had never used the Internet at home offered this explanation.

Internet Access at Other Locations

Three percent of households without Internet at home cited the ability to go online at locations other than their homes. African Americans and Asian Americans were slightly more likely to do so, with 4 percent of each group reporting they could use the Internet elsewhere, than white and Hispanic households at 3 percent each.

More than any other group, non-using households headed by someone with a college degree or higher could use the Internet elsewhere but not at home (7 percent), compared to 1 percent of such householders without a high school diploma. Heads of households between 16 and 44 years old were six times more likely than those ages 65 and older to cite the ability to go online at other locations as the main reason for not using the Internet at home (6 percent compared to 1 percent). Similarly, 6 percent of employed non-using householders reported they could use the Internet elsewhere, compared to 1 percent of householders not in the labor force. Non-adopting households with school-age children were more likely than those without them to cite the ability to use the Internet at other locations (5 percent compared to 3 percent).

Lack of Service Availability

One percent of non-using householders responded they did not access the Internet at home because service was not available in their area, while the figure was 2 percent for rural residents. Twenty-one percent of dial-up users stated they did not use broadband at home because high-speed Internet service was unavailable in their area, which was the third most popular response among this group behind expense and no need or interest in the service.

Privacy or Security Concerns

Only 1 percent of non-using households expressed concerns about privacy or security, which was the least common reason for not using the Internet at home in 2011.

CONCLUSION

The Internet has become integral to daily life in the United States, as users increasingly depend on continuous access to a wide range of online capabilities. Looking at Americans' online activities for the first time in eight years revealed some interesting developments. In October 2003, CPS data showed that personal communications, such as email, and general information were the most popular activities; by July 2011, financial services and entertainment had also become very common online activities. Focusing on 2011, unemployed Internet users proved to be nearly twice as likely to search for work online as their employed counterparts. On the other hand, the historically more- connected young, those more highly educated, and urban dwellers were more likely to research healthcare information online than older persons, the less educated, and rural residents. Mobility in communications and reliance on the Internet in daily life are increasingly common, led by younger Americans, but growing in importance to older generations. Internet use outside the home helped meet the needs of unconnected households. For those who have not yet adopted the Internet, digital literacy training and low-cost Internet service options are helping ensure that affordable broadband is available to all Americans and that they are skilled users.

In light of the growing range of valuable online activities, individuals who lack connectivity find themselves at a significant disadvantage compared with their peers. Today more than nine out of ten households are located in areas with at least some level of high-speed Internet service available. However, broadband adoption continues to lag availability; currently about seven out of ten households nationally have online connectivity, and there is variability among demographic groups and geographic locations.

Regarding adoption, disparities persist as home broadband use by low-income, less-educated, African American, Hispanic, senior, disabled, or rural households lagged behind other groups and the national average. However, there is good news regarding a narrowing "digital divide" as demand for broadband increases, particularly among at least some groups that in the past had been among the least connected. For example, householders age 65 and older experienced an increase in broadband adoption from 2010 to 2011 of 4 percentage points (45 percent to 49 percent). Broadband use among households headed by someone with a disability also saw an increase in adoption during that time period, specifically 3 percentage points (from 43 percent to 46 percent). During the same nine- month span, rural households in aggregate gained connectivity by one point (from 57 percent to 58 percent).

On the other hand, in July 2011, some 30 percent of the 119 million American households did not use broadband at home. The three main reasons for non-use of the Internet at home reported in the 2011 CPS survey include "don't need it/not interested" (48 percent), "too expensive" (28 percent), and "no computer/computer inadequate" (13 percent). Variations on these percentages occurred among those who never used or formerly used the Internet at home (e.g., former home Internet- using households valued cost more highly). Overall among non-using households, a lack of service availability was not one of the key obstacles to using the Internet at home except for one group: Twenty-one percent of dial-up users stated they did not use broadband at home because it was not available in their area.

Continued growth of the Internet economy requires policymakers to understand the nature and extent of users' online activities, and armed with this knowledge, craft policies that prompt broadband availability and adoption. The Department of Commerce will continue reporting such developments through ongoing data collection and analysis, and will continue to implement policies and initiatives that promote Americans' continued ability to benefit from engaging online with friends, employers, medical professionals, and many others.

APPENDIX A. DATA AND METHODOLOGY

This report uses data from the July 2011 Current Population Survey ("CPS"), a monthly survey of a representative sample of the U.S. noninstitutional population that provides data on labor force participation, income, and demographic characteristics of households. In addition, this report analyzes data from the most recent (July 2011) CPS Computer and Internet Use Supplement, a special supplement to the CPS that periodically gathers information on Internet use.

The July 2011 Computer and Internet Use Supplement asked each surveyed household whether someone in that household used or owned a computer, as well as who in the household used the Internet, and the devices and locations from which they did so (home, office, school, library, and other places). In addition, the survey asked the household which of the following technologies members utilized to connect to the Internet from home: dial-up service, DSL, cable modem, fiber optics, satellite, mobile broadband, or some other Internet connection technology. Using these data, it is possible to determine whether a household owned a personal computer (i.e., desktop, laptop, netbook, or notebook) or a handheld device, as well as the type of broadband technology (i.e., DSL, cable modem, fiber optics, satellite, or mobile broadband) members utilized to connect to the Internet. The survey also asked those households that did not access Internet services to state their main reason for not doing so.

New to the Computer and Internet Use supplement this year was a group of questions primarily concerning Internet usage habits.[24] These questions were asked of the household respondent about his/her Internet and computer use. Since one person in each household responded and that person was not randomly selected, a special set of weights was created to properly tally these variables. Care should be used when analyzing these data because the respondents are not evenly distributed across the sample based on age. This is why Section 2 only examines the population ages 25 and older.

About 53,500 household records comprise our sample, representing 119.3 million American households. We analyze computer and Internet use at the household and person levels and their association with characteristics such as age, family income, household size and composition, and geographic location. When conducting household-level analysis, we use information for the household as proxies for characteristics like education, race, ethnicity, age, disability status, and foreign-born status. In this report we use the words "adoption," "use," "utilization," and "access" interchangeably to indicate that a household reported having Internet access.

Prior to October 2010, data on computer use, as well as the types of broadband technology that online households utilized, have not been available since the early 2000s. The supply and demand for both mobile devices and residential Internet access services have changed enormously during this period. The July 2011 data serve to update these trends. The data from the July 2011 CPS make it possible to continue to identify the preferred or most common types of computers and access technologies used for residential Internet access, including the prevalence of mobile broadband technologies and handheld devices.

APPENDIX B. STATISTICAL TABLES

Table B1. Marginal Effects from Regression of Employment on Internet Use and Demographic and Geographic Characteristics, Persons Age 25+, 2011

Variable	Coefficient Estimate	Standard Error
Internet User	0.0588 **	0.0050
Internet at Home	0.0282 **	0.0054
Female	-0.0284 **	0.0056
Married, Spouse Present	0.0778 **	0.0061
Married, Spouse Absent	0.0389 *	0.0157

Variable	Coefficient Estimate	Standard Error
Widowed	0.0209 *	0.0084
Divorced	0.0328 **	0.0061
Separated	0.0210 †	0.0119
Female and Married, Spouse Present	-0.1584 **	0.0070
Education: High School Diploma	0.0585 **	0.0057
Education: Some College	0.0831 **	0.0061
Education: College Degree or More	0.1401 **	0.0057
Age	0.0220 **	0.0006
Age Squared	-0.0003 **	0.0000
Race: African American	-0.0499 **	0.0059
Race: Hispanic	0.0231 **	0.0056
Race: Asian American	-0.0263 **	0.0082
Race: Other	-0.0412 **	0.0120
Others Employed in Household	0.1000 **	0.0049
Household Size	-0.0299 **	0.0028
Related School-Age Children at Home	0.0340 *	0.0135
Household Size When Related School-Age Children at Home	0.0008	0.0037
Lives in Owned Home	0.0294 **	0.0045
Disabled	-0.2655 **	0.0050
Foreign-Born Non-Citizen	-0.0006	0.0070
Metropolitan Status: Urban	-0.0048	0.0043
Metropolitan Status: Unidentified	0.0244	0.0160
Region: Midwest	0.0041	0.0048
Region: South	-0.0039	0.0044
Region: West	-0.0236 **	0.0057
Constant	0.2540 **	0.0182
Sample Size	89,498	
Estimated Age 25+ Population Size	201,474,711	
R^2	0.3046	

† indicates 90 percent confidence that the marginal effect is nonzero ($p < 0.1$).

* indicates 95 percent confidence that the marginal effect is nonzero ($p < 0.05$).

** indicates 99 percent confidence that the marginal effect is nonzero ($p < 0.01$).

Note: Data for whites, African Americans, and Asian Americans do not include people of Hispanic origin. Persons of Hispanic origin may be of any race.

Table B2. Marginal Effects from Regression of Discussing News with Others on Internet Use, Media Use, and Demographic and Geographic Characteristics, Persons Age 25+, 2011

Variable	Coefficient Estimate	Standard Error
Internet User	-0.0004	0.0070
Internet at Home	-0.0196 **	0.0070
News Sources: Social Media	0 1293**	0.0074
News Sources: Internet	0.2018 **	0.0057
News Sources: Television	0.0708**	0.0061
News Sources: Radio	0.2751 **	0.0052
News Sources: Print Media (Newspapers, Magazines, etc.)	0.2419**	0.0049
Female	0.0339 **	0.0041
Family Income: $25,000-49,999	0.0004	0.0057
Family Income: $50,000-74,999	- 0.0021	0.0068
Family Income: $75,000-99,999	0.0051	0.0081
Family Income: $100,000 or More	- 0.0058	0.0076
Education: High School Diploma	- 0.0280 **	0.0073
Education: Some College	- 0.0131 †	0.0079
Education: College Degree or More	- 0.0132	0.0083
Age	- 0.0016 †	0.0008
Age Squared	0.0000	0.0000
Race: African American	- 0.0277 **	0.0067
Race: Hispanic	- 00309	0.0076
Race: Asian American	- 0.0164	0.0112
Race: Other	0-0134	0.0146
Household Size	- 0.0044 †	0.0025
Related School Age Child e at Home	0.0251 †	0.0146
Household Size When Related School-Age Children at Home	0.0066 †	0.0039
Disabled	0.0167 **	0.0062
Foreign-Born Non-Citizen	- 0.0227 *	0.0092
Metropolitan Status Urban	- 0.0150**	0.0053
Metropolitan Status: Unidentified	- 0.0016	0.0213
Constant	0.1120**	0.0241
Sample Size	50,371	
Estimated Age 25+ Population Size	201,474,710	
R^2	0.3469	

† indicates 90 percent confidence that the marginal effect is non zero ($p < 0.1$).

* indicates 95 percent confidence that the marginal effect is nonzero ($p < 0.05$).
** indicates 99 percent confidence that the marginal effect is nonzero ($p < 0.01$).
Note: Data for whites, African Americans, and Asian Americans do not include people of Hispanic origin. Persons of Hispanic origin may be of any race.

Table B3. Home Broadband and Dial-Up Internet Use by State, Percent of Households, 2011

State	Total Households	Broadband Adoption			Dial-Up Only		
		Percent	95% Confidence Interval		Percent	95% Confidence Interval	
			Lower Bound	Upper Bound		Lower Bound	Upper Bound
Alabama	1,917,329	59.3	53.1	65.3	4.5	3.1	6.5
Alaska	266,446	73.1	68.7	77.2	3.6	2.3	5.6
Arizona	2,634,043	72.1	68.2	75.8	2.3	1.4	3.9
Arkansas	1,166,405	59.9	54.5	65.1	2.3	1.5	3.6
California	12,949,021	72.6	71.2	73.9	1.9	1.5	2.3
Colorado	1,966,436	78.7	74.9	82.1	1.4	0.7	2.5
Connecticut	1,338,113	76.5	73.8	79.1	1.5	1.0	2.2
Delaware	342,274	70.4	66.3	74.1	2.1	1.2	3.5
District of Columbia	282,881	70.1	67.0	73.1	0.9	0.5	1.6
Florida	7,831,939	73.5	71.5	75.5	1.3	0.9	2.1
Georgia	3,757,808	67.0	63.5	70.2	3.1	2.4	4.1
Hawaii	438,771	70.0	65.5	74.1	1.6	0.9	2.9
Idaho	571,357	72.2	66.3	77.5	2.6	1.3	4.9
Illinois	4,921,452	70.8	68.2	73.2	1.8	1.3	2.6
Indiana	2,553,331	64.6	61.0	68.0	3.3	2.3	4.7
Iowa	1,203,115	68.1	65.1	70.9	3.6	2.2	5.7
Kansas	1,118,341	72.7	69.3	76.0	2.4	1.5	3.9
Kentucky	1,745,782	62.3	57.2	67.1	3.4	1.8	6.3
Louisiana	1,764,817	62.8	57.8	67.6	2.7	1.2	5.8
Maine	540,319	70.9	67.8	73.8	3.8	2.7	5.3
Maryland	2,121,727	75.7	73.0	78.2	1.2	0.7	2.0
Massachusetts	2,564,338	76.5	73.4	79.3	0.9	0.5	1.8
Michigan	3,899,829	68.7	66.2	71.1	4.0	3.1	5.3
Minnesota	2,107,215	76.0	73.5	78.2	2.5	1.9	3.4

Table B3. (Continued)

State	Total Households	Broadband Adoption			Dial-Up Only		
		Percent	95% Confidence Interval		Percent	95% Confidence Interval	
			Lower Bound	Upper Bound		Lower Bound	Upper Bound
Mississippi	1,128,947	53.5	47.4	59.4	2.9	1.3	6.3
Missouri	2,459,661	63.6	60.3	66.8	2.4	1.6	3.7
Montana	440,274	65.6	60.2	70.7	1.7	0.7	3.8
Nebraska	714,897	69.1	66.1	72.0	2.1	1.5	3.0
Nevada	1,037,353	74.5	71.1	77.7	0.9	0.4	2.0
New Hampshire	530,119	79.2	76.6	81.5	3.1	2.2	4.2
New Jersey	3,233,204	74.8	71.8	77.6	2.2	1.3	3.6
New Mexico	794,101	58.3	52.6	63.8	2.0	1.2	3.4
New York	7,644,991	70.8	68.5	72.9	2.1	1.6	2.9
North Carolina	3,789,718	66.8	63.5	70.0	1.6	1.0	2.6
North Dakota	282,530	70.2	65.8	74.2	0.6	0.2	1.6
Ohio	4,702,772	66.1	62.6	69.5	2.4	1.8	3.2
Oklahoma	1,481,923	62.8	57.0	68.3	3.1	2.1	4.4
Oregon	1,498,068	76.5	73.3	79.5	2.0	1.3	3.3
Pennsylvania	5,027,347	67.9	65.5	70.2	1.8	1.2	2.7
Rhode Island	418,352	71.8	68.6	74.8	1.6	1.0	2.6
South Carolina	1,798,695	62.2	58.9	65.4	2.6	1.5	4.3
South Dakota	323,989	68.3	65.0	71.5	2.7	1.8	4.1
Tennessee	2,649,281	61.4	57.6	65.1	2.2	1.3	3.7
Texas	9,156,263	64.3	62.2	66.3	2.9	2.2	3.8
Utah	912,549	77.0	72.8	80.8	1.7	1.0	3.1
Vermont	269,441	70.4	67.0	73.5	3.9	2.8	5.5
Virginia	2,960,611	68.3	65.4	71.1	2.5	1.7	3.7
Washington	2,704,273	79.5	75.8	82.8	2.4	1.6	3.6
West Virginia	753,384	59.1	54.4	63.6	3.2	2.0	5.1
Wisconsin	2,288,041	72.5	69.0	75.7	2.1	1.3	3.4
Wyoming	231,521	72.4	69.0	75.5	1.8	0.9	3.4

Table B4. Household Computer Ownership by State, Percent of Households, 2011

State	Total Households	Computer Ownership		
		Percent	95% Confidence Interval	
			Lower Bound	Upper Bound
Alabama	1,917,329	69.2	63.5	74.5
Alaska	266,446	81.0	77.2	84.2
Arizona	2,634,043	78.2	74.7	81.4
Arkansas	1,166,405	67.7	63.1	72.0
California	12,949,021	78.4	77.0	79.8
Colorado	1,966,436	82.5	78.1	86.2
Connecticut	1,338,113	81.3	78.5	83.7
Delaware	342,274	75.3	71.7	78.5
District of Columbia	282,881	74.2	71.0	77.1
Florida	7,831,939	78.6	76.7	80.3
Georgia	3,757,808	75.5	72.4	78.4
Hawaii	438,771	72.9	68.6	76.9
Idaho	571,357	81.8	76.3	86.3
Illinois	4,921,452	76.3	74.2	78.1
Indiana	2,553,331	70.9	67.4	74.1
Iowa	1,203,115	76.0	72.9	78.8
Kansas	1,118,341	78.3	74.9	81.3
Kentucky	1,745,782	71.2	67.1	74.9
Louisiana	1,764,817	69.5	65.1	73.6
Maine	540,319	76.9	74.1	79.4
Maryland	2,121,727	80.0	77.5	82.3
Massachusetts	2,564,338	78.8	75.9	81.5
Michigan	3,899,829	75.7	73.1	78.1
Minnesota	2,107,215	81.4	79.4	83.3
Mississippi	1,128,947	64.5	58.9	69.8
Missouri	2,459,661	71.6	68.4	74.7
Montana	440,274	74.1	67.5	79.7
Nebraska	714,897	77.3	73.3	80.9
Nevada	1,037,353	77.6	74.4	80.5
New Hampshire	530,119	84.2	81.6	86.5
New Jersey	3,233,204	78.9	76.2	81.3
New Mexico	794,101	68.5	63.5	73.1

Table B4. (Continued)

State	Total Households	Computer Ownership		
		Percent	95% Confidence Interval	
			Lower Bound	Upper Bound
New York	7,644,991	75.3	73.3	77.3
North Carolina	3,789,718	72.6	69.7	75.3
North Dakota	282,530	76.3	73.7	78.7
Ohio	4,702,772	71.8	68.5	74.9
Oklahoma	1,481,923	72.1	66.0	77.5
Oregon	1,498,068	82.2	79.5	84.6
Pennsylvania	5,027,347	74.6	72.4	76.7
Rhode Island	418,352	77.3	74.4	79.9
South Carolina	1,798,695	68.5	65.5	71.3
South Dakota	323,989	75.7	71.9	79.1
Tennessee	2,649,281	69.5	65.6	73.1
Texas	9,156,263	71.5	69.6	73.4
Utah	912,549	84.4	80.2	87.8
Vermont	269,441	79.3	76.4	82.0
Virginia	2,960,611	74.6	71.6	77.4
Washington	2,704,273	84.9	82.1	87.3
West Virginia	753,384	64.6	60.7	68.2
Wisconsin	2,288,041	77.0	74.1	79.7
Wyoming	231,521	79.0	76.2	81.5

Table B5. Home Broadband Use in Urban and Rural Areas by State, Percent of Households, 2011

State	Broadband Adoption in Urban Areas				Broadband Adoption in Rural Areas			
	Urban Households	Percent	95% Confidence Interval		Rural Households	Percent	95% Confidence Interval	
			Lower Bound	Upper Bound			Lower Bound	Upper Bound
Alabama	1,426,339	65.7	60.3	70.6	490,990	40.9	25.3	58.6
Alaska	183,790	74.1	68.3	79.1	82,656	71.0	62.1	78.5
Arizona	2,330,074	74.4	70.3	78.1	303,969	54.6	39.8	68.5
Arkansas	776,048	65.0	59.4	70.2	390,357	49.8	41.9	57.7

State	Broadband Adoption in Urban Areas				Broadband Adoption in Rural Areas			
	Urban Households	Percent	95% Confidence Interval		Rural Households	Percent	95% Confidence Interval	
			Lower Bound	Upper Bound			Lower Bound	Upper Bound
California	12,659,938	72.8	71.4	74.1	289,082	65.6	59.2	71.4
Colorado	1,740,811	79.3	75.6	82.5	*			
Connecticut	1,284,181	76.7	74.0	79.2	53,932	72.3	53.8	85.3
Delaware	271,377	70.6	66.1	74.8	70,898	69.4	61.2	76.5
District of Columbia	282,881	70.1	67.0	73.1	**			
Florida	7,545,094	74.1	71.9	76.1	286,845	60.1	41.5	76.2
Georgia	3,187,875	70.8	67.0	74.3	569,933	45.3	39.5	51.2
Hawaii	305,722	70.7	66.0	75.1	133,048	68.2	58.2	76.7
Idaho	363,221	78.0	72.6	82.6	208,137	62.1	48.8	73.9
Illinois	4,336,566	72.3	69.7	74.7	584,886	59.5	51.8	66.9
Indiana	1,874,030	66.5	62.7	70.2	679,301	59.2	49.6	68.1
Iowa	711,472	72.5	67.3	77.2	491,643	61.6	57.1	66.0
Kansas	710,434	75.4	70.7	79.5	407,907	68.2	60.8	74.8
Kentucky	954,119	69.1	63.2	74.3	791,663	54.1	45.2	62.7
Louisiana	1,479,206	62.7	58.1	67.0	*			
Maine	273,051	74.2	69.7	78.3	267,269	67.5	63.0	71.6
Maryland	2,030,390	76.4	73.6	79.0	91,337	59.4	53.3	65.2
Massachusetts	2,492,427	76.2	73.0	79.0	*			
Michigan	3,280,129	70.2	67.3	72.9	619,700	60.9	56.0	65.7
Minnesota	1,553,179	78.5	75.4	81.3	554,036	68.9	65.8	71.9
Mississippi	484,158	59.1	51.8	66.1	644,790	49.2	40.9	57.6
Missouri	1,917,479	66.1	61.8	70.1	542,182	54.9	47.7	61.9
Montana	153,980	70.7	63.2	77.2	286,294	62.9	55.4	69.9
Nebraska	436,936	75.5	71.8	78.9	277,961	59.1	53.0	65.0
Nevada	897,711	75.8	72.2	79.1	*			
New Hampshire	323,525	81.5	78.2	84.5	206,594	75.5	71.2	79.3
New Jersey	3,233,204	74.8	71.8	77.6	*			
New Mexico	563,670	63.5	57.0	69.6	230,431	45.5	29.9	62.1
New York	6,938,860	72.5	70.1	74.8	706,131	53.5	47.1	59.8
North Carolina	2,559,400	72.7	69.3	75.9	1,230,318	54.6	47.7	61.2
North Dakota	146,074	74.5	69.3	79.2	136,456	65.5	58.6	71.9
Ohio	3,595,497	68.6	65.4	71.6	1,107,276	58.2	47.5	68.1
Oklahoma	1,021,850	68.5	63.4	73.3	460,073	50.2	43.4	56.9

Table B5. (Continued)

State	Broadband Adoption in Urban Areas				Broadband Adoption in Rural Areas			
	Urban Households	Percent	95% Confidence Interval		Rural Households	Percent	95% Confidence Interval	
			Lower Bound	Upper Bound			Lower Bound	Upper Bound
Oregon	1,149,854	78.5	74.2	82.2	348,215	70.1	66.0	73.8
Pennsylvania	4,086,389	70.0	67.3	72.5	940,958	58.9	52.3	65.1
Rhode Island	418,352	71.8	68.6	74.8	*			
South Carolina	1,167,988	66.2	61.3	70.8	630,707	54.7	50.2	59.1
South Dakota	159,526	73.3	67.9	78.2	164,463	63.4	59.1	67.6
Tennessee	1,977,161	63.6	58.6	68.4	672,120	54.8	48.3	61.1
Texas	8,134,534	65.4	63.3	67.4	1,021,729	55.8	45.4	65.7
Utah	699,587	77.8	73.0	82.0	*			
Vermont	81,233	85.9	80.6	89.9	188,209	63.7	59.4	67.7
Virginia	2,489,932	73.0	69.5	76.1	470,680	43.5	31.9	55.9
Washington	2,507,665	79.9	75.4	83.8	196,609	74.5	62.8	83.5
West Virginia	435,647	65.0	59.0	70.5	317,738	51.0	44.7	57.4
Wisconsin	1,690,756	75.0	71.1	78.6	597,285	65.3	59.3	70.8
Wyoming	71,230	73.1	67.0	78.4	160,291	72.0	68.2	75.6

* Rural broadband adoption estimates for the following states were not included due to insufficient data for rural areas in the CPS: Colorado, Louisiana, Massachusetts, Nevada, New Jersey, Rhode Island, and Utah.

** The District of Columbia had no rural areas.

REFERENCES

American Telemedicine Association (ATA). (2012). Telemedicine frequently asked questions (FAQs). Retrieved from http://www.americantelemed.org/learn/what-is-telemedicine/faqs

Bausman, K. & Goe, W. R. (2004). An examination of the link between employment volatility and the spatial distribution of property crime rates. *The American Journal of Economics and Sociology, 63*(3), 665-695. Retrieved from http://search.proquest.com/docview/217675407

Belo, R. & Ferreira, P. (2012). Spillover effects of broadband in schools and the critical role of children. *2012 TPRC*. doi:10.2139/ssrn.2032097

Brynjolfsson, E., Yu (Jeffrey) Hu & Smith, M. D. (2003). Consumer surplus in the digital economy: Estimating the value of increased product variety at online booksellers. *Management Science, 49*(11), 1580-1596. Retrieved from http://search.proquest.com/docview/213234771

Bureau of Labor Statistics (BLS). (2009). How the government measures unemployment. Retrieved from http://www.bls.gov/cps/cps_htgm.htm

Choo, S., Mokhtarian, P. L. & Salomon, I. (2005). Does telecommuting reduce vehicle-miles traveled? An aggregate time series analysis for the U.S. *Transportation, 32*(1), 37-64. doi:10.1007/s11116-004-3046-7

Crandall, R., Lehr, W. & Litan, R. (2007). The effects of broadband deployment on output and employment: A cross-sectional analysis of U.S. data. *Issues in Economic Policy*, (6). Retrieved from http://www.brookings.edu/views/papers/crandall/200706litan.pdf

Economics and Statistics Administration (ESA) & National Telecommunications and Information Administration (NTIA). (2011). *Exploring the digital nation: Computer and Internet use at home*. Retrieved from http://www.esa.doc.gov/Reports/exploring-digital-nation-computer- and-internet-use-home

Federal Communications Commission (FCC). (2010). *Schools and libraries universal service support mechanism and a national broadband plan for our future* (CC Docket No. 02-6, GN Docket No. 09-51, FCC 10-175). Retrieved from http://www.gpo.gov/fdsys/pkg/FR-2010-12-03/pdf/2010-29386.pdf

Federal Communications Commission (FCC). (2011). *Telephone subscribership in the United States*. Retrieved from http://www.fcc.gov/document/fcc-releases-telephone-subscribership-report-0

Federal Communications Commission (FCC). (2012a). Lifeline reforms saved more than $210 million in 2012, beating target; FCC announces 14 pilot projects to increase broadband adoption using a portion of savings from Lifeline reform [Press release]. Retrieved from http://transition.fcc.gov/Daily_Releases/Daily_Business/2012/db1219/DOC-318059A1.pdf

Federal Communications Commission (FCC). (2012b). *Ninth broadband progress notice of inquiry* (GN Docket No. 12-228, FCC 12-91, ¶ 3). Retrieved from http://www.fcc.gov/document/ fcc-launches-ninth-inquiry-broadband-availability

Gillett, S., Lehr, W., Osorio, C. & Sirbu, M. A. (2006). Measuring broadband's economic impact. *MIT Communications Futures Program*. Retrieved from http://cfp.mit.edu/publications/CFP_Papers/Measuring_bb_econ_impact-final.pdf

Hale, T. (2012, August 16). Is there such a thing as an online health lifestyle? Proceedings from the American Sociological Association Annual Meeting. Denver, CO.

Harrison, J. P. & Lee, A. (2006). The role of e-Health in the changing health care environment. *Nursing Economics*, *24*(6), 283-288. Retrieved from https://www.nursingeconomics.net/ce/2008/article12283289.pdf

International Telecommunication Union (ITU). (2011). Share of Internet users in the total population. In *The World in 2011: ICT Facts and Figures*. Retrieved from http://www.itu.int/ITU-D/ict/facts/2011/material/ICTFactsFigures2011.pdf

Jennings, M. K. & Zeitner, V. (2003). Internet use and civic engagement: A longitudinal analysis. *Public Opinion Quarterly*, *67*(3), 311-334. doi:10.1086/376947

Moy, P., Manosevitch, E., Stamm, K. & Dunsmore, K. (2005). Linking dimensions of Internet use and civic engagement. *Journalism and Mass Communication Quarterly*, 82(3), 571-586. doi:10.1177/107769900508200306

National Telecommunications and Information Administration (NTIA). (2011). *Digital nation: Expanding Internet usage*. Retrieved from http://www.ntia.doc. gov/report/2011/digital- nation-expanding-internet-usage-ntia-research-preview

National Telecommunications and Information Administration (NTIA) & Federal Communications Commission (FCC). (2013). Broadband statistics report: Access to broadband technology by speed. *National Broadband Map* (June 2012 data). Retrieved from http://www.broadbandmap.gov/download/Technology%20by%20Speed.pdf

The Nielsen Company. (2012). America's new mobile majority: A look at smartphone owners in the U.S. *nielsenwire*. Retrieved from http://blog.nielsen.com/nielsenwire/?p=31688

Office for National Statistics. (2012, August 24). Internet access – Households and individuals, 2012. Retrieved from http://www.ons.gov.uk/ons/rel/rdit2/internet-access---households-and- individuals/2012/index.html

Office of Management and Budget (OMB). (2009). *Update of statistical area definitions and guidance on their uses* (OMB Bulletin No. 10-02). Retrieved from http://www.whitehouse.gov/sites/ default/files/omb/assets/bulletins/b10-02.pdf

Pew Internet & American Life Project (Pew Internet). (2012, March 1). Nearly half of American adults are smartphone owners. Retrieved from http://pewinternet.org/Reports/2012/ Smartphone-Update-2012/Findings.aspx

Putnam, R. D. (2000). *Bowling alone: The collapse and revival of American community*. New York, NY: Simon & Schuster.

Rains, S. A. (2008). Health at high speed: Broadband Internet access, health communication, and the digital divide. *Communication Research*, *35*(3), 283-297. doi:10.1177/0093650208315958

Romm, J. (2002). The Internet and the new energy economy. *Resources, Conversation and Recycling*, 36(3), 197-210. doi:10.1016/S0921-3449(02)00084-8

Smidt, C. (1999). Religion and civic engagement: A comparative analysis. *Annals of the American Academy of Political and Social Science*, *565*(1), 176-192. doi:10.1177/000271629956500112

Stenberg, P., Morehart, M., Vogel, S., Breneman, V., Cromartie, J. & Brown, D. (2009). *Broadband Internet's value for rural America.* Retrieved from http://www.ers.usda.gov/publications/err-economic-research-report/err 78.aspx

Tolbert, C. J. & McNeal, R. S. (2003). Unraveling the effects of the Internet on political participation? *Political Research Quarterly*, 56(2), 175-185. doi:10.1177/106591290305600206

U.S. Census Bureau. (2012). Poverty thresholds by size of family and number of children (2011). Retrieved from http://www.census.gov/hhes/www/poverty/data/threshld/index.html

The White House, Office of the Press Secretary. (2011, February 10). *Remarks by the President on the national wireless initiative in Marquette, Michigan*. Retrieved from http://www.whitehouse. gov/the-press-office/2011/02/10/remarks-president-national-wireless-initiative-marquette-michigan

The White House, Office of the Press Secretary. (2012, June 13). We can't wait: President Obama signs executive order to make broadband construction faster and cheaper [Press release]. Retrieved from http://www.whitehouse.gov/the-press-office/2012/06/13/we-can-t-wait-president-obama-signs-executive-order-make-broadband-const

End Notes

[1] In this report, as in previous analyses, ESA and NTIA examine residential Internet service – service that connects homes to the Internet – from the demand side based on the Census

Bureau's survey of households. The terms "adoption," "use," "utilization," "access," and "connection" are used interchangeably to indicate that a household reported having Internet service. The term "Internet service" includes both the provision of dial-up service and broadband service. Similarly, the CPS survey inquired about households' ownership or use of a home computer to examine whether they have available to them the devices people used to go online. The report describes such home computer access as "ownership" or "use" and employs the terms interchangeably. This study employs the terms "usage" and "online activities" in a similar way.

[2] Jennings and Zeitner (2003) determined that Internet use may increase levels of civic engagement (or at least abate a generally negative trend). Tolbert and McNeal (2003) found that Internet users were more likely to vote in the 1996 and 2000 presidential elections.

[3] The National Broadband Map ("NBM") provides detailed broadband availability data. Launched in 2011, the map displays data collected semiannually by NTIA's State Broadband Initiative ("SBI") in collaboration with the Federal Communications Commission ("FCC") and 56 state and territorial partners. The NBM database provides a snapshot of the United States on a very granular (Census block) level, yielding substantial geographic data relating to broadband availability. Data available from the map include each company providing broadband service to a Census block, the technology used to provide each service, and the maximum advertised speed of the service (both download and upload).

[4] See Appendix A for a detailed description of the data and methodology employed in this report.

[5] Americans are far ahead of the world as a whole, with only one-third of the world's 1.8 billion households online in 2011 (ITU, 2011).

[6] CPS Supplement questions about computer use have evolved over time to keep up with the increasing computing capabilities of mobile devices. In 2003, the survey asked only about the presence of desktop and laptop personal computers in a household; however, in 2010 the increasing popularity of mobile devices prompted the inclusion of "handheld computers," including smartphones, in the questionnaire. Staff further refined the methodology in 2011 by specifically asking about the presence of "desktop, laptop, netbook, notebook, and tablet computers" in a household, while a separate question addressed mobile phone usage. Due to the highly portable nature of mobile phones, the 2011 CPS Supplement gathered data on usage at any location rather than strictly at home.

[7] This observation applies outside the United States as well. For example, in the United Kingdom, 80 percent of all households had Internet service at home in 2012, including 95 percent of households with children (Office for National Statistics, 2012).

[8] ESA and NTIA have tested all comparisons referenced in this paper, which are significant at the 95 percent confidence level.

[9] We use the terms "employed," "unemployed," and "not in the labor force" in the same fashion as the Bureau of Labor Statistics. For example, a person who is unemployed has no job, is currently available for work, and either has actively looked for work during the past four weeks or is expecting to be recalled from temporary layoff (BLS, 2009).

[10] "Fully employed persons" refers to employed persons who are not considered to be underemployed. Underemployed persons have a job, but work fewer than 35 hours per week and want to work full-time.

[11] For example, the FCC noted in December 2012 that "over 80 percent of Fortune 500 companies… require online job applications" (FCC, 2012a).

[12] Data for whites, African Americans, Asian Americans, and American Indian and Alaska Natives do not include people of Hispanic origin, who may be of any race.

[13] For example, the odds of employment rose by 4 percentage points between ages 25 and 36 and then gradually declined every year after. Women who have never married were around 3 percentage points less likely to be employed than similarly-situated males, while married women were over 18 percentage points less likely to be employed compared with married men. Higher levels of education were also associated with increased likelihood of

employment, with college graduates having been nearly 14 percentage points more likely to be employed than adults lacking a high school diploma. Persons with disabilities were approximately 27 percentage points less likely to be employed than their peers without disabilities. Finally, despite controlling for a wide range of factors, the model predicts that race and ethnicity still correlated with employment outcomes. Relative to whites, African Americans were nearly 5 percentage points less likely to be employed.

[14] For example, Hale (2012) suggests that "the Internet has the potential to alleviate social disparities in health by providing greater access to health information and other health-related resources."

[15] The geographic variable for identifying a household's location as urban or rural is not available in the CPS public use files. This report uses the terms "urban" and "rural" to refer to metropolitan and nonmetropolitan areas, respectively. The definition of a metropolitan area (effective since 2000) is based on "core based statistical area" ("CBSA"), which includes both metropolitan and micropolitan statistical areas. According to the 2000 standards, each CBSA must have at least one urban area with at least 10,000 inhabitants. Each metropolitan statistical area must contain at least one urbanized area with a population of 50,000 or more. Each micropolitan statistical area must contain at least one urban cluster with a population of between 10,000 and 50,000 (OMB, 2009).

[16] Some experts also refer to these activities as examples of "telehealth." The American Telemedicine Association considers "telehealth" and "telemedicine" to be synonymous, while acknowledging that others in the field "have parsed out unique definitions for each word" (ATA, 2012).

[17] Income had no statistically significant impact on this aspect of civic engagement, and the predicted effects of education were small and only partially significant. A few factors exhibited clearer relationships with discussing the news; women were 3 percentage points more likely than men to discuss current events when holding other variables constant, disabled persons were 2 percentage points more likely to do so than their non-disabled counterparts, and non-citizens were 2 percentage points less likely than citizens to engage in this activity. The comparatively muted observed effects based on demographic factors highlight the importance of general media attentiveness in predicting whether an individual discussed current events with others. However, some of those factors – especially education – might have correlated with an individual's propensity to obtain news information from any source, so the model may underestimate their importance.

[18] For example, academic research on civic engagement suggests that regular attendance at religious services and other religiously affiliated activities may increase community involvement (Smidt, 1999). There may also be other unknown variables omitted from the model that affect the likelihood of discussing current events.

[19] A household with at least one of the following high-speed, high capacity, two-way Internet services is considered to have broadband: DSL, cable modem, fiber optics, satellite, mobile broadband, or some other non-dial-up Internet connection. The CPS did not ask about the speed of the particular broadband service a household uses because of the difficulty of gathering the information. The household-level estimate on broadband Internet access based on the total sample has a margin of error of plus or minus 0.46 percentage points.

[20] Note that Figure 16 refers only to mobile broadband adoption **at home**. The higher mobile broadband adoption rates reported in some other studies measure activities that are not strictly limited to the use of a mobile broadband device at home.

[21] States are ordered by estimated average household broadband adoption rate for ease of understanding and not as a specific ranking. Rates for broadband, dial-up, and computer use should be understood in the context of their associated confidence intervals, set forth in Appendix B, Tables B3 and B4.

[22] The following states were not included in this discussion due to insufficient data for rural areas in the CPS: Colorado, Louisiana, Massachusetts, Nevada, New Jersey, Rhode Island, and Utah. The District of Columbia had no rural areas. Estimated adoption rates should be

understood in the context of their associated confidence intervals, set forth in Appendix B, Table B5.

[23] If all broadband technologies are included that approximate the FCC's speed threshold for broadband at 3 megabits per second download and 768 kbps upload, including mobile broadband, then the coverage increases to 98 percent of the population (NTIA & FCC, 2013).

[24] The specific questions used in the July 2011 survey instrument are available at http://www.census.gov/prod/techdoc/ cps/cpsjul11.pdf.

In: Assessing the Digital Nation
Editor: Darius L. Shea
ISBN: 978-1-62948-362-7

Chapter 2

FOUR YEARS OF BROADBAND GROWTH*

Office of Science and Technology Policy &
The National Economic Council

INTRODUCTION

Broadband access is an essential part of our economy. Given the reliance of so many American businesses and families on this basic technology, it's easy to overlook that just 15 years ago broadband barely existed for consumers.

To create jobs and grow wages at home, and to compete in the global information economy, the delivery of fast, affordable and reliable broadband service to all corners of the United States must be a national imperative. The build-out of broadband infrastructure itself is a major driver of American investment and job creation, but even more significant are the ways that connectivity is transforming a range of industries, from education to entertainment to agriculture to travel. High-speed wired and wireless networks place the United States at the center of a digital economy that is one of the brightest parts of our short-term recovery and long-term competitiveness.

This report highlights (1) the status of broadband development in the United States in 2013, including both progress made in achieving greater

---

* This is an edited, reformatted and augmented version of the Office of Science and Technology Policy & The National Economic Council, dated June 2013.

availability to broadband as well as the significant remaining challenges of timely and affordable deployment of broadband to all Americans; (2) several important contributions of the Obama Administration over the last four years that have accelerated the growth of broadband technology; and (3) two key opportunities for further action.

I. THE STATE OF AMERICAN BROADBAND IN 2013

By nearly any metric the last four years have been a period of tremendous growth in broadband infrastructure, access, and the digital economy upon which they rely.

Since the President took office, national broadband availability has increased at all advertised speed levels.[1] Today, about 91 percent of Americans have access to wired broadband speeds of at least 10 Mbps downstream, and 81 percent of Americans have access to similarly fast mobile wireless broadband.

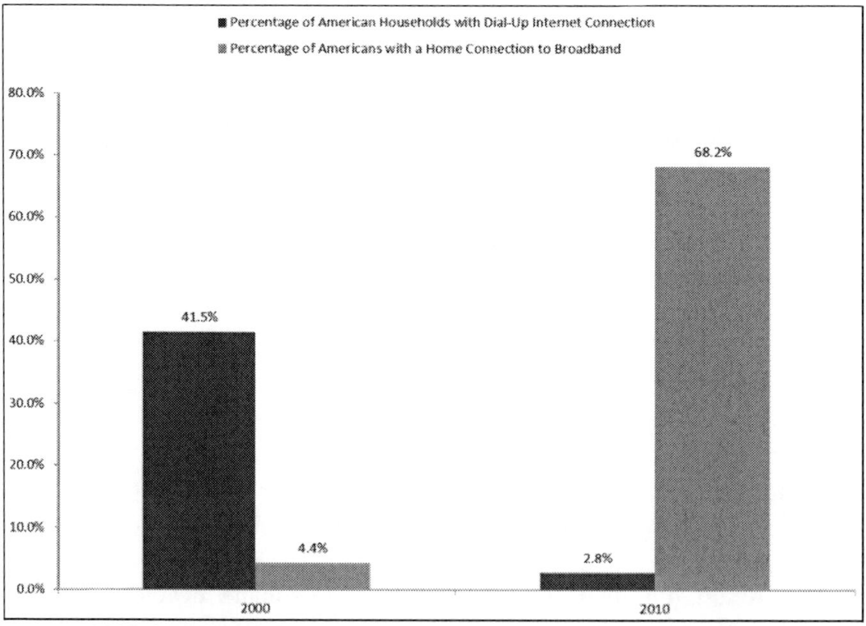

Note: In 2000 broadband was considered to be 200 kbps while dial-up internet connections were typically 28.8 bps or 56 kbps.

Figure 1. The Transformation in American Connectivity[8].

The last four years have also been a period of significant transition in the needs of many broadband subscribers. Taking advantage of low barriers to entry and an open network, providers of cloud computing, remote processing, media-rich content and streaming video have created unprecedented demand for higher-bandwidth Internet services to businesses and households alike. That demand has driven substantial investment in high-speed facilities.

The government commonly defines the "basic" speed for broadband at 3 Mbps downstream and 768 kbps upstream (3Mb/768kb), with some regulatory decisions defining basic service as 4 Mbps downstream and 1 Mbps upstream.[2] Nonetheless, we acknowledge that the country is rapidly reaching the point at which baseline broadband evaluations should increase, and might instead begin at 10 Mbps downstream.[3] This evolving baseline reflects a growing need for increased bandwidth as more Americans use the Internet for work and to build career skills.[4]

Changing attitudes about broadband also track with the growth in demand and availability. The number of broadband adopters continues to increase steadily, alongside the rising number of Americans who consider the Internet a valuable learning tool. In 2011, 77 percent of American Internet users ages 25 and older reported relying on the Internet for personal communications, while 66 percent relied on it for general information – and about half depended on the Internet for financial services, for consumer services and for entertainment.[5]

While broadband access and speed have grown significantly over the last four years, significant areas for improvement remain, as the FCC has documented.[6] Moreover, many markets are subject to at most limited competition among broadband providers.[7] Thus, areas for improvement remain. Ongoing efforts are needed to ensure that all Americans have access to affordable and reliable high-speed access – and that the infrastructure exists to support increases in consumer and business demand for higher speeds.

Significant Private Investment

Responding to the increasing consumer demand for services accessed through broadband, the private sector has been driving important advances in infrastructure and technology. U.S. telecommunications firms have made significant investments in infrastructure; for example, just two of the largest U.S. telecommunications companies account for greater combined stateside investment than the top five oil/gas companies, and nearly four times more than the big three auto companies combined.[9] In fact, since President Obama took office in early 2009, nearly $250 billion in private capital has been invested in U.S. wired and wireless broadband networks. In just the last two years, more high-speed fiber cables have been laid in the United States than in any similar period since 2000.[10] Moreover, during President Obama's first term, the annual investment in U.S. wireless networks alone grew more than 40 percent from $21 billion to $30 billion. Projections for 2013 estimate an annual wireless network investment at $35 billion.[11]

These investments in broadband, along with the openness of the network, are part of a self-perpetuating cycle that coupled with increased broadband adoption and digital literacy skills training, have significantly improved access to higher-speed wireline and wireless broadband services, brought new users online and enabled the "App Economy," a phenomenon that itself has driven demand for and deployment of even faster high-speed services.

High-Speed Wired and Wireless Access for More Americans

Evaluating the United States' progress in broadband access requires a careful weighing of a number of characteristics, including access, speed, and geography. Measured against these factors, digital networks in the United States have improved greatly from even just a few years ago, reaching a larger percentage of households and offering much faster speeds.

Fast Speeds. The figure below displays the most recent broadband-speed availability data from the National Telecommunications and Information Administration (NTIA).

	≥ 3 Mbps	≥ 6	≥ 10	≥25	≥ 50	≥ 100	≥ 1 Gbps
All Broadband	98.18%	96.17%	94.39%	78.51%	75.15%	47.09%	3.17%
Wireline	93.41%	92.81%	90.91%	78.11%	74.85%	46.87%	3.17%
Wireless	94.37%	84.17%	80.66%	4.94%	3.03%	1.80%	0.00%

Figure 2. Percentage of U.S. Population with Access to Various Advertised Broadband Download Speeds (Mbps)[12].

	≥ 3 Mbps / 768 kbps	≥ 6	≥ 10	≥25	≥ 50	≥ 100	≥ 1 Gbps
Cable	86.92%	86.95%	86.15%	76.42%	72.63%	44.20%	0.00%
DSL	73.51%	64.60%	47.39%	7.21%	0.11%	0.01%	0.00%
Fiber	20.20%	20.00%	19.86%	18.72%	18.25%	6.79%	3.16%
Fixed	34.33%	25.81%	10.89%	4.88%	2.99%	1.78%	0.00%
Mobile	91.81%	80.58%	78.67%	0.00%	0.00%	0.00%	0.00%
Copper	43.25%	15.37%	14.59%	1.46%	0.27%	0.12%	0.01%

Figure 3. Percentage of Population with Access to Broadband Speeds by Technology Type[13].

Notably, wireless (mobile and fixed) broadband has demonstrated remarkable gains in availability, and wired broadband access continues to improve. Even with 93% availability, however, millions of Americans remain without access to fixed broadband, which often provides the most robust connections. As described below, the Commerce Department and the Federal Communications Commission have made progress, and are taking numerous important steps to address this challenge through a number of ongoing programs.

Leading Wireless Technology. With the deployment of fourth-generation (4G) technologies such as Long-Term Evolution (LTE), advanced mobile broadband service is also much more widely available today than it was two years ago.[14] The most recently available data from NTIA (June 2012) indicates that 80.6 percent of Americans now have access to advanced mobile broadband.

To place those figures in context, not only was the United States the first country to deploy 4G LTE networks at scale, but also, as of mid-2012 nearly half the world's LTE subscribers were within the United States.[15] Likewise, in 2012, North America's average mobile data connection speed was 2.6 Mbps – the fastest such average in the world, nearly twice that available in Western Europe, and over five times the global average.[16]

The average connection speed in the United States in the fourth quarter of 2012 was 7.4 Mbps, the eighth fastest among all nations, and the fastest when compared to other countries with either a similar population or land mass.[17]

Platform for Innovation, Growth and Consumer Benefits: Smartphones, Tablets and other Internet-Connected Devices

As of the first quarter of 2013, 500 million Internet-connected devices were in American homes.[18] These include smartphones, tablets, and other Internet-connected devices that run applications. Between 2009 and 2012, U.S. smartphone ownership increased from 16 percent to 56 percent.[19] More than 80 percent of smartphones sold globally in 2012 run operating systems developed by U.S. companies, up from 5 percent six years ago and 25 percent three years ago. Moreover, a growing number of households have transitioned away from wired phone lines, and the number with only mobile phones today exceeds 35 percent.[20]

These devices have done more than connect Americans to one another more easily. The integration of mobile broadband, advanced operating systems and increasingly sophisticated hardware, along with low barriers to entry to an open network, have enabled an entire economy of mobile applications to develop in the United States. This "App Economy" is one of America's most dynamic and growing sectors, and one that industry studies have cited as creating more than 500,000 U.S. jobs since 2007.[21]

Economic studies have also placed the "consumer surplus" or value consumers gain from their time online at roughly 2 percent of their income – a particularly impressive outcome given that Americans spend a tiny fraction of that amount to get broadband service.[22] It is easy to imagine why: another study found that in answering everyday questions, those who had access to a search engine did so in fifteen fewer minutes than those without Internet access.[23]

The benefits are not just for the individual and household. Broadband is known to make a substantial contribution to national economic growth, as more and more citizens become empowered to take advantage of innovative technologies and connect to each other in new ways. Subsequent study of that consumer surplus suggests that Internet access contributed to the economy an average of $34 billion per year since 2002 (equivalent to about 0.26 percent to annual GDP growth).[24] A separate study of 25 OECD countries between 1996 and 2007 found that a 10 percent increase in broadband penetration raises per-capita GDP growth by 0.9-1.5 percentage points. Another study found that within 66 high income countries during the period 1980-2002 a 10 percent increase in broadband penetration yielded an additional 1.21 percentage points of GDP growth.[25] Increasing the speed of a nation's broadband connectivity also provides an economic boost – a study of 33 OECD countries between 2008 and 2010 concluded that doubling the broadband speed for an economy increases GDP by 0.3 percent.[26]

Significant Challenges Remain: Adoption, Speed, and Pricing

Today, more than 56 percent of Americans have a smartphone,[27] and more than 78 percent of Americans are Internet users. However, 72 percent of Americans use broadband at home. Despite the advances in broadband coverage – particularly considering the United States' size and geography – adoption is lower than in some nations with comparable Gross Domestic Product (GDP) per capita.[28] Connecting more homes in the United States remains an important policy priority; absent this connectivity, individuals cannot take advantage of opportunities for education and skills development, entrepreneurship and e-commerce available to those with the connectivity.

Given the promise that broadband connections hold for households, businesses, and the economy as a whole, low adoption, availability of higher-speed connections, and affordability remain concerns for national broadband policy.

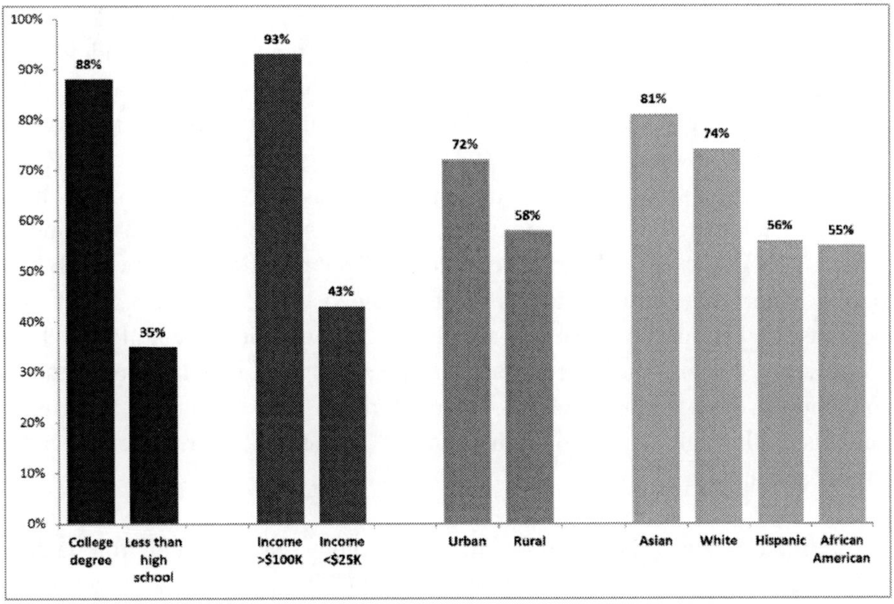

Figure 4. Percentage of Broadband Adoption by Demographic Group[30].

Uneven adoption. Demographic factors influence broadband adoption. For example, home broadband adoption among those with at least a college degree (88%) is more than double that of those who did not complete high school (35%); 50 percentage points separate broadband use of households with annual incomes exceeding $100,000 from those with incomes below $25,000 (93% compared to 43%). A larger percentage of urban dwellers use broadband at home (72%) than rural residents (58%), while Asians (81%) and Whites (74%) have adopted broadband at home to a greater extent than Hispanics (56%) and African Americans (55%).[29]

In total, approximately 29 percent of Americans do not use broadband at home as of 2011 – this includes 8 percent of Americans who access the Internet only away from home (for instance at the workplace or an anchor institution such as a library or community center) and less than 3 percent of Americans who continue to use dial-up at home. Why do millions of these Americans who have access still choose not to subscribe? According to recent surveys, cost, skills and relevance remain major concerns.

Lack of Interest. In 2011, almost half of all households that did not use the Internet at home reported they had no need or interest in going online. This was especially true for older Americans; 68 percent of Americans 65 years or older were likely to express no need or interest in using the Internet at home, compared to 21 percent of people between the ages of 16 and 44.

Those who frequently visit the Internet report using it for a variety of purposes. In 2011, users at least 25 years of age stated they rely on the Internet for: personal communications (77%); general information (66%); entertainment (48%); job searches, job training, or working from home (29%); healthcare (27%); and education or schoolwork (23%), among others.[31] NTIA's Broadband Technology Opportunities Program (BTOP) public computer center projects are helping to teach non-users about these benefits to encourage them to adopt this empowering technology. These surveys are corroborated by international data suggesting that broadband use in schools drives home adoption; in Portugal, for instance, students in schools with high- speed broadband were 20 percent more likely to adopt broadband at home.[32]

Affordability. The second most common reason cited for lack of broadband adoption is affordability. In 2011, 28 percent of non-adopting households did not have broadband service at home primarily because of the expense of Internet service, an increase of four percentage points above

2010 data. Cost represents a significantly greater obstacle for lower income Americans: compared to 18 percent of households with annual incomes above $50,000 that cited affordability as the main deterrent to going online, 32 percent of families with incomes below $25,000 responded that the high cost of Internet service prevented them from using broadband at home.

The cost of broadband varies significantly based on speed, type of service package, and geographical location; the data about those costs also varies significantly. Based on a limited sample of prices available on provider websites, the FCC reports that in 2011 dollars, the average monthly price for a 1-5 Mbps connection was $35, the average cost of a 5-15 Mbps connection was $44, and the average price of a 15-25 Mbps connection was $56.50.[33] The cost of "triple-play" packages (Internet, TV, and phone) also vary. New America Foundation found that plan prices in the United States started at $66/month for a basic package that included broadband speeds of 6 Mbps downstream and 1 Mbps upstream as well as TV and phone service. At the higher end, symmetrical service at 30 Mbps coupled with phone and TV was $130/month. For a stand-alone wireline service, price varies significantly depending on location and provider. At around $35, for example, a customer in San Francisco can subscribe to a 200 Mbps symmetrical connection; New York and Washington, D.C., residents can subscribe to a 25 Mbps down/2 Mbps up service; and citizens of Los Angeles can purchase a 10 Mbps down/1 Mbps up service.[34] Most data indicate the price consumers pay has remained steady or increased year-to-year, even though speed is increasing for those base prices over time; thus price remains one of the key challenges to increasing broadband adoption in the United States.

	≥25 Mbps	≥ 50	≥ 100	≥ 1 Gbps
Cable	76.42%	72.63%	44.20%	0.00%
DSL	7.21%	0.11%	0.01%	0.00%
Fiber	18.72%	18.25%	6.79%	3.16%
Fixed Wireless	4.88%	2.99%	1.78%	0.00%
Mobile	0.00%	0.00%	0.00%	0.00%
Copper	1.46%	0.27%	0.12%	0.01%

Figure 5. Percentage of Population with Access to Top-Tier Broadband Speeds[36].

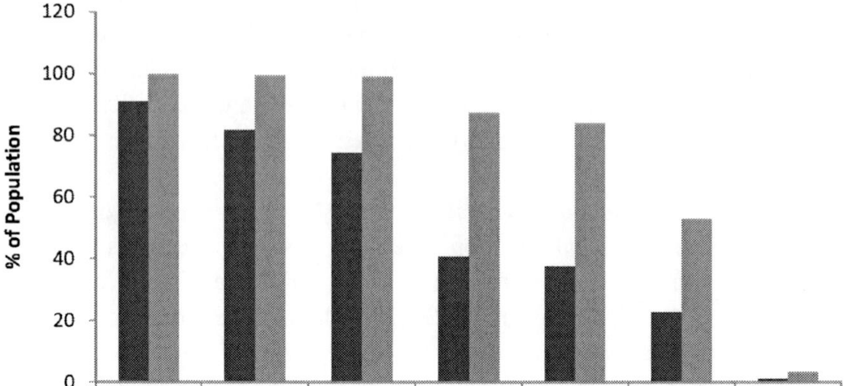

	≥ 3/768 Mbps	≥ 6 Mbps	≥ 10 Mbps	≥ 25 Mbps	≥ 50 Mbps	≥ 100 Mbps	≥ 1 Gbps
Rural	91.1	81.8	74.3	40.8	37.6	22.8	1.2
Urban	99.9	99.6	99.1	87.5	84.0	52.9	3.6

Figure 6. Broadband Availability by Urban and Rural (June 2012)[37].

In addition to increasing adoption rates, the United States must also prepare for the increasing need for faster speeds by consumers, businesses and community anchor institutions. While broadband is accessible through multiple technologies at lower speeds, far fewer technologies offer the faster speeds necessary to take full advantage of emerging innovations (see below).[35] For example, while cable, DSL, and mobile wireless are all widely available at basic broadband speeds, at 25 Mbps, only cable (76.42%) is widely available, followed next by fiber (18.72%).

Disparity between Urban and Rural Access

Of great concern is also the disparity between urban and rural access, as well as uneven distribution of high-speed access across certain underserved geographies. Today, while almost 100 percent of urban residents have access to download speeds of at least 6 Mbps, only about 82 percent of residents in rural communities can access those speeds. The disparity is even more pronounced at higher speeds: almost 88 percent of urban residents have access to speeds of 25 Mbps, but less than half that percentage, about 41 percent, have access to those speeds in rural communities. Of course not all users require those higher speeds, and satellite and terrestrial wireless technologies continue to deliver promising improvements. But even considering the higher costs often associated with providing broadband to rural areas, limits in availability should not hold back rural populations seeking higher-speed, high quality Internet access.

Access to broadband in rural communities also differs considerably by state. In 10 states and four territories, 20 percent or less of the rural population has access to speeds of 25 Mbps. In 15 states and Puerto Rico, between 21 percent and 40 percent of the rural population has access to speeds of 25 Mbps. In only 12 states do more than 60 percent of their rural populations have access to at least this speed.[38]

The tables below highlight those states with the greatest broadband access at the middle three tiers of available speeds, as well as those with the least such access.

While it is a positive development that users often have the opportunity to choose among wired and wireless services at basic speeds, high-speed connections continue to show room for improvement both in availability, cost, and competition, particularly as more homes, companies and community institutions require higher speeds.

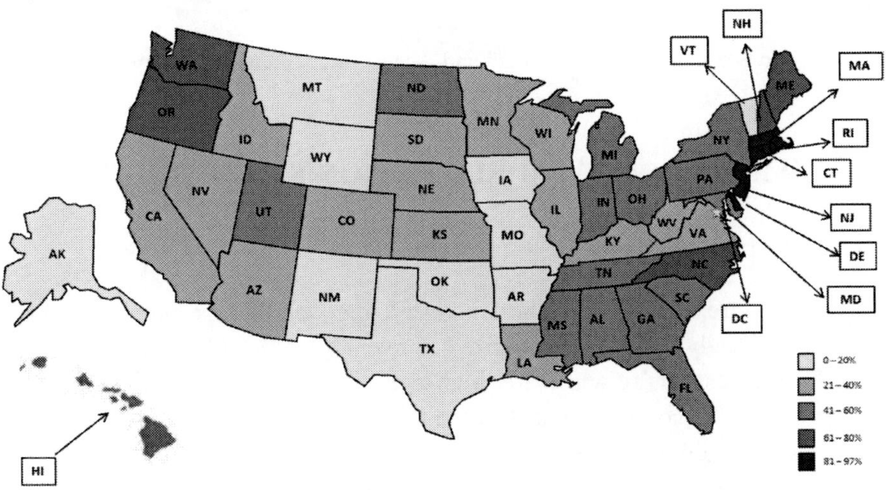

Figure 7. Percent of Population in Rural Areas with Access to 25 Mbps Speeds[39].

Most Access: ≥ 10 Mbps			Most Access: ≥ 25 Mbps			Most Access: ≥ 50 Mbps		
1	DC	100%	1	Rhode Island	99.62%	1	Rhode Island	99.62%
2	Rhode Island	99.80%	2	Connecticut	99.09%	2	DC	98.27%
3	Connecticut	99.75%	3	DC	98.44%	3	Connecticut	97.08%
4	New Jersey	99.48%	4	New Jersey	97.77%	4	Hawaii	96.94%
5	Delaware	99.29%	5	Hawaii	96.94%	5	Massachusetts	96.55%
6	Massachusetts	99.16%	6	Massachusetts	96.84%	6	New Jersey	96.37%
7	Maryland	98.99%	7	Delaware	95.81%	7	Delaware	95.78%
8	Hawaii	98.70%	8	Washington	95.28%	8	Washington	94.76%
9	Florida	98.47%	9	Oregon	92.73%	9	Oregon	92.20%
10	New York	98.39%	10	California	91.22%	10	New York	90.97%
11	Utah	98.35%	11	New York	91.04%	11	California	89.56%
12	Washington	98.15%	12	Florida	90.71%	12	Utah	89.22%
13	California	97.95%	13	Utah	90.71%	13	Maryland	88.91%
14	Pennsylvania	97.15%	14	Maryland	89.46%	14	Nevada	84.33%
15	Illinois	96.90%	15	Illinois	87.53%	15	Michigan	83.05%

Figure 8. 15 States with Most Broadband Access at 10, 25 and 50 Mbps[40].

Least Access: ≥ 10 Mbps			Least Access: ≥ 25 Mbps			Least Access: ≥ 50 Mbps		
47	Wyoming	81.58%	47	Arkansas	28.54%	47	Iowa	19.81%
48	Alaska	81.40%	48	Wyoming	22.17%	48	New Mexico	10.85%
49	Mississippi	78.57%	49	Vermont	20.48%	49	Wyoming	2.27%
50	Montana	75.52%	50	Montana	12.71%	50	Montana	1.42%
51	West Virginia	69.69%	51	Alaska	0.06%	51	Alaska	0.00%

Figure 9. Five States with Least Broadband Access at 10, 25 and 50 Mbps.

II. The Obama Administration's Contributions

Key Points

- More than $2.8 billion invested in over 220 projects that have already built and improved over 90,000 miles of broadband infrastructure and made high-speed connections available to about 14,000 community institutions.
- Created tax incentives to spur one of the largest private-sector infrastructure investments in history.
- On track to ensure that at least 98% of Americans have access to 4G wireless broadband service by 2016, and to free up 500 MHz of wireless spectrum for commercial use.

> Once we emerge from the immediate crisis, the long-term economic gains to communities that have been left behind in the digital age will be immeasurable. Studies have shown that when communities adopt broadband access, it can lead to hundreds of thousands of new jobs. Broadband can remove geographic barriers between patients and their doctors. It can connect our kids to the digital skills and 21st century education required for the jobs of the future. And it can prepare America to run on clean energy by helping us upgrade to a smarter, stronger, more secure electrical grid.
>
> - President Barack Obama, July 2, 2010

The Obama Administration has made a number of contributions to the rapid growth of high- speed broadband, including (1) stoking broadband investment through the Recovery Act; (2) providing tax incentives for business investment; (3) freeing up spectrum for mobile broadband; and (4) a light-touch, multi-stakeholder approach to regulation that has fostered both innovation in applications and deployment of infrastructure.

Recovery Act Investment in Broadband

One important goal of the Recovery Act was to increase access to and drive adoption of broadband across America, including an important $6.9 billion upgrade to the nation's broadband infrastructure. As part of the appropriation, $4.4 billion was administered by the Department of Commerce's NTIA to deploy broadband infrastructure, support public computer centers and encourage adoption of broadband. The remaining $2.5 billion was directed to the Department of Agriculture's Rural Utilities Service (RUS) to expand broadband access in rural areas.

Under the RUS's Broadband Initiatives Program, over $1.4 billion has been awarded to 105 broadband projects in 37 states and one territory. NTIA's Broadband Technology Opportunities Program (BTOP) received more than 2,800 applications requesting more than $36 billion in support for potential projects in all 50 states and territories. Grant recipients were required to show that "but for" these Federal grant funds, there would not be a business case to serve these communities. As of the end of May 2013, NTIA had invested over $2.8 billion in over 220 projects in every state, territory and the District of Columbia, the vast majority of which has been spent building and improving over 90,000 miles of broadband infrastructure and making high-speed connections available to about 14,000 community institutions across the country.[41]

BTOP grants fell into three categories:

1. ***Comprehensive Community Infrastructure.*** Projects to deploy new or improved broadband facilities (e.g., laying new fiber-optic cables or upgrading wireless towers) and to connect "community anchor institutions" including schools, libraries, hospitals, and public safety facilities. These networks help ensure sustainable community growth and provide the foundation for enhanced household and business broadband Internet services.
2. ***Public Computer Centers.*** Projects to establish new public computer facilities or upgrade existing ones that provide broadband access to the general public or to specific vulnerable populations, such as low-income individuals, the unemployed, seniors, children, minorities, and people with disabilities.
3. ***Sustainable Broadband Adoption.*** Projects that focus on increasing broadband Internet usage and adoption, including among vulnerable populations where broadband technology traditionally has been underutilized. Many projects include digital literacy training and outreach campaigns to increase the relevance of broadband in people's everyday lives.

Additionally, through the State Broadband Initiative (SBI), NTIA awarded approximately $293 million to 56 recipients, one grant each to the 50 states, five territories, and the District of Columbia. With this funding, states have been collecting and validating data biannually on the availability, speed, type, and location of broadband services, as well as the broadband services used by community anchor institutions, such as schools, libraries, and hospitals. The result of this effort is a pioneering data source, the National Broadband Map (www.broadbandmap.gov) that has enabled the Administration to provide the best quality information on broadband services ever available in the United States. In addition, states are also using this grant funding to organize around broadband capacity building, using the data to drive policy and support local and regional efforts to compete in the digital economy.

These Recovery Act investments are helping to improve access to education and healthcare services as well as boost economic development for communities stymied by limited or no access to high-speed broadband.

The investments in broadband infrastructure, public computer centers, and sustainable adoption have helped provide job training to the unemployed or under-employed, helped school children access the materials they need to learn, allowed rural doctors to connect to more specialized medical centers, and enabled small businesses to offer their services to national and international markets. BTOP projects, for example, have already delivered about 10 million hours of technology training to about 2.8 million users. Important lessons about how to overcome barriers to adoption are captured in **NTIA's Broadband Adoption Toolkit** – a further return on the nation's investment.[42]

BTOP accompanied another important development in expanding broadband access: the Federal Communications Commission's *Connect America Fund.* After a decade of unsuccessful reform attempts, the FCC voted unanimously in October 2011 to overhaul the Universal Service Fund (USF).[43] The USF was originally designed as a program for delivering telephone service to rural America; its reform created the Connect America Fund to ensure broadband access to all Americans. The Connect America Fund is a public-private partnership that will over 10 years work to extend broadband to un-served consumers and communities across the country. As part of its overhaul, the Commission also established access to mobile broadband as a universal service goal for the first time, and created the Mobility Fund to spur deployment of advanced wireless infrastructure and eliminate dead zones. In their first phases of implementation, the Connect America Fund and Mobility Fund are already helping extend home broadband to nearly 400,000 previously un-served Americans and bring mobile broadband to up to 83,000 previously-uncovered road miles.

Tax Incentives for Private Investment

Businesses making capital investments in broadband, like in other areas, are allowed to recover the cost of those expenditures over time through a depreciation expense deduction. By allowing an immediate deduction (or "expensing") of investment costs, the tax relief that the President has signed into law lowers the effective tax rate on income derived from business investments. (Expensing provisions reduce the after-tax cost of investment by allowing companies to deduct the cost of their investment in the year it purchases qualified tangible property and places that property into service, rather than stretching that deduction out based on the depreciation schedule for that investment.) This incentive increases the effective rate of return on investment, leading companies to make investments that might otherwise be unattractive.

The President signed into law one of the largest temporary tax incentives for investment in American history, which has played a key role in accelerating and increasing investment across the economy – and particularly in telecommunications. In addition to this incentive expensing that was in effect in 2011, the President also signed into law additional investment incentives every year since 2009.

Furthermore, by providing an immediate window of opportunity in which the costs of investing in qualified property are lower, a temporary expensing policy encourages firms to accelerate investment that they might delay for several years – an especially powerful incentive for investments in longer-lived assets, like broadband infrastructure.

As part of the 2009 Recovery Act, the President allowed businesses to recover the costs of certain capital expenditures made in 2008 and 2009 – including capital investments in broadband infrastructure – faster than under the ordinary schedule, by depreciating 50 percent of the cost in the year the investor placed the property in service. The President's 2010 Small Business Jobs Act extended this first-year 50 percent bonus depreciation for qualifying property purchased and placed in service in 2010.[44]

In December 2010, the President signed into law legislation that provided immediate 100 percent expensing of investment costs, one of the largest temporary investment incentives in American history, which allowed businesses and investors to deduct immediately the full cost of broadband investments rather than having to depreciate them over time – a benefit that Treasury estimated could encourage $50 billion in total additional investment.[45] The American Taxpayer Relief Act provided 50 percent bonus depreciation for investments made in 2013.[46]

This tax incentive had an especially significant effect on the broadband industry. In a joint 2011 letter, leaders from Comcast and Time Warner described the effect expensing had on their ability to invest in broadband infrastructure: "Despite the downturn in the economy, the cable communications sector has been able to continue steady investment and to retain jobs as a result of policies like 100 percent expensing."[47] And in 2010, the Independent Telephone & Telecommunications Alliance (ITTA) released a statement commending Congress for passing the President's Small Business Jobs Act "and specifically for the inclusion of bonus depreciation language included in the Act … [which] will allow ITTA member companies to accelerate broadband investment and deliver the economic, educational, and healthcare opportunities that broadband provides."[48]

Freeing Spectrum for New Innovation

As the President stated in June 2010, "America's future competitiveness and global technology leadership depend, in part, upon the availability of additional spectrum. The world is going wireless and we must not fall behind." The need for spectrum is a constant refrain, and the Administration continues to devote significant attention to the issue.

The public policy imperative was highlighted by the Council of Economic Advisors that, in its February 2012 report, concluded, "it is unlikely that wireless carriers will be able to accommodate this surging demand without additional spectrum...In short, the projected growth in data traffic can be achieved only by making more spectrum available for wireless use."[49] Private-sector studies have reached similar conclusions, noting that wireless data traffic in North America is likely to increase 100-fold between 2009 and 2016.[50]

To address this need, in 2010 the President directed NTIA to work with the FCC to re-purpose 500 MHz of spectrum from existing Federal or nonfederal use to wireless broadband within 10 years, almost doubling the amount of spectrum available for that use. The challenge to NTIA and the agencies was a daunting one: within the spectrum bands best suited for wireless broadband service, dozens of Federal agencies operate thousands of spectrum-dependent systems supporting national defense, law enforcement, and severe weather forecasting, among other vital Federal missions. Yet within just a few months, NTIA had identified 115 MHz of federally-assigned spectrum that could be re-purposed for broadband and described its plan and methodology for reviewing additional spectrum bands to reach the 500 MHz goal. In executing upon that plan, NTIA supported the formation of working groups made up of representatives of Federal agencies, commercial broadband providers, and other stakeholders who are engaged in unprecedented collaborative efforts to free up more spectrum for the FCC to allocate to wireless broadband service while safeguarding critical Federal activities.

In parallel with these efforts, the President announced in his 2011 State of the Union address a commitment to ensure that at least 98 percent of

Americans had access to 4G wireless broadband service by 2016. Thanks to private and public investments resulting from spectrum policy, tax incentives, Recovery Act investments, and competition policy, we are on track to achieve the goals of bringing 4G to nearly all Americans by the President's deadline.

Central to the President's wireless agenda are the spectrum provisions he proposed in the American Jobs Act, which were carried forward in the Middle Class Tax Relief and Job Creation Act of 2012. This legislation expanded the FCC's authority to auction spectrum, enabling more robust wireless broadband service for consumers while also raising billions of dollars in revenue. The most innovative feature of the bill is the grant of authority to the FCC to design and conduct "incentive auctions," in which television broadcasters can voluntarily bid on the right to relinquish some or all of their spectrum rights in return for some of the proceeds raised when that spectrum is auctioned for flexible use, such as wireless broadband. The FCC has proposed rules for the incentive auctions and is reviewing public comments now.

Country	Current	Pipeline	Current + Pipeline
USA	608	55+	663+
Australia	478	230	708
Brazil	554	0	554
China	227	360	587
France	555	50	605
Germany	615	0	615
Italy	540	20	560
Japan	500	10	510
Spain	540	60	600
U.K.	353	265	618

Note: US pipeline numbers do not include the significant amount of spectrum that will be made available for mobile broadband from incentive auctions and federal repurposing.

Figure 10. Summary of Total Available Licensed Spectrum Available for Mobile Broadband (in megahertz).

In addition to providing incentives for federal agencies sharing or relocating from spectrum reallocated to commercial users, last year's legislation also established the First Responders Network Authority, or FirstNet, an independent authority within NTIA tasked with the design and deployment of a nationwide interoperable wireless network for first responders. FirstNet will empower regional, state, local, tribal, and territorial public safety agencies with the advantages of nationwide scale and interoperability across boundaries, while also giving them local control and flexibility. At the same time, with dedicated spectrum of its own and $7 billion in funding from spectrum auctions, FirstNet will partner with commercial providers and other stakeholders to expand wireless broadband access across the nation.

USA  EU 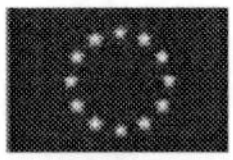

Band	USA Current	USA Pipeline	EU Current	EU Pipeline (Unknown)
TV White Spaces [Unl1]	0-150	+	-	
863-870 MHz	-	-	7[Unl2]	
902-928 MHz	26[Unl3]	-	-	
1880-1930 MHz	10[Unl4]	-	20[Unl5]	
2400-2483.5 MHz[Unl6]	83.5	-	83.5	
3550-3700 MHz	50[Unl7]	100[Unl8]	-	
5150-5350 & 5470-5825 MHz[Unl9]	555	-	555	
5350-5470 & 5850-5925 MHz	-	195[Unl10]	-	
TOTAL	724.5-874.5	295+	665.5	

Figure 11. Amount of Spectrum Below 6 GHz Available for Unlicensed Broadband Use.

In comparison to other nations, the United States ranks among the top countries in current licensed spectrum available for mobile broadband, and while we cannot predict the amount of spectrum freed by the incentive auctions, this forthcoming spectrum — combined with Federal repurposing — are likely to keep the United States well atop other nations in mobile broadband allocation.

Also vital to our economy is unlicensed spectrum, which is responsible for many of the innovative wireless technologies in devices we use every day, such as Wi-Fi. Last year's legislation confirmed the FCC's discretion to designate spectrum for unlicensed use, such as Wi- Fi. Unlicensed spectrum plays an important role in fostering new wireless uses by enabling "innovation without permission." Unlicensed spectrum applications have transformed the personal electronics industry, and are poised to make substantial contributions to the retail, manufacturing, and other sectors. Additionally, low power, close-range connections like Wi-Fi are essential to easing the strain on the cellular networks of the major wireless broadband carriers. Studies indicate that the majority of traffic over smartphones and tablets is now carried over Wi-Fi.[51] In comparison to the European Union, the United States has substantially more spectrum (as well as more diverse spectrum) available for unlicensed applications.

Light-Touch, Multi-Stakeholder Policies to Build Opportunity and Trust

The Administration has also taken a comprehensive approach to broadband and Internet policy, with a series of policies, multi-stakeholder efforts, and processes that are improving the efficiency and predictability of the digital economy. They include:

- **FCC's 2010 Open Internet Order.** In 2010 the FCC established light-touch rules designed to ensure that the Internet remains an open platform for innovation and expression, and that content is delivered regardless of its origin or recipient. The rules ensure low barriers to entry by providers of Internet-based services, which in turn continues to stimulate demand for service and facilities by making the Internet more valuable to users.
- **Consumer Privacy Bill of Rights.** Because trust in digital systems and the companies handling personal information is essential to the

economy, the Administration released the nation's first-ever comprehensive policy document on digital privacy. As part of that policy's commitment to encouraging innovation, minimizing regulatory burden, and engaging all stakeholders, NTIA has for the last year led an inclusive, voluntary process for developing privacy codes of conduct.

- **National Strategy for Trusted Identities in Cyberspace.** To help promote an ecosystem with better security than the standard username and password, the White House released the National Strategy for Trusted Identities in Cyberspace (NSTIC), and has staffed a national program office at the National Institutes of Standards and Technology (NIST) to coordinate the Federal activities needed to implement the trusted identity ecosystem. That office has been leading efforts to promote private-sector involvement, build consensus on legal and policy frameworks, work with industry to identify new standards, and promote pilot projects.
- **Cybersecurity.** In addition to the nation's first top-to-bottom review of national cybersecurity policy (the Cyberspace Policy Review), President Obama has also taken important steps to reduce the economic drag and uncertainty caused by cybersecurity vulnerabilities. Specifically, he outlined a series of legislative actions in his May 2011 Legislative Proposal, and in 2013 signed Executive Order 13636, which expanded information sharing, developed robust privacy protections, and set up an inclusive, multi- stakeholder effort to develop nation-wide standards for critical infrastructure cybersecurity.
- **International Strategy for Cyberspace.** The U.S. International Strategy for Cyberspace is the United States' comprehensive foreign policy document for the digital world, outlining our vision for sustaining a global online environment that is open to innovation, interoperable the world over, and secure and reliable enough to support our economic growth. The strategy explains how the multi-stakeholder model, free data flows, respect for Internet Freedom, and strong regulatory forbearance helps enhance prosperity globally.
- **Internet Policy Board.** To regularize the development of national policy related to the digital economy, the White House also established in early 2013 the Internet Policy Board, a senior-level forum for policy development and decision-making. The Board includes regular membership from the President's Cabinet agencies,

as well as several components of the Executive Office of the President.

III. Policy Opportunities

Key Points

- **Schools and community anchors** remain underserved, particularly at connection speeds that would support important advances in education crucial to national competitiveness.
- **Spectrum sharing** will be crucial to the future of wireless broadband, and the federal government should encourage the policies and technology advances to enhance it.

There are a variety of opportunities to drive the broadband agenda forward, including important priorities like promoting competition, continuing to fuel investment and innovation, and protecting consumers. This report focuses on two important areas where policy innovation can help unlock the full potential of broadband technology for Americans: expanding high- speed access for schools and anchor institutions, and continuing efforts to relieve the spectrum crunch.

Schools and Anchor Institutions

> "[I]n an age when the world's information is a just click away, it demands that we bring our schools and libraries into the 21st century. We can't be stuck in the 19th century when we're living in a 21st century economy. That's why, today, we're going to take a new step to make sure that virtually every child in America's classrooms has access to the fastest Internet and the most cutting-edge learning tools. And that step will better prepare our children for the jobs and challenges of the future and it will provide them a surer path into the middle class. And, as a consequence, it will mean a stronger, more secure economy for all of us."
>
> - President Obama, Mooresville, North Carolina, June 6, 2013

Driven by new digital technologies, the future of learning is increasingly interactive, individualized, and full of real-world experiences and information. While broadband access has improved considerably in recent years, broadband infrastructure in America's K-12 schools and local libraries has lagged far behind. Unfortunately, despite the successes of the FCC's E-Rate program connecting schools to the Internet for the first time, oversubscription means that the average school has about the same connectivity than the average American home, but serves 200 times as many users, and fewer than 20 percent of educators say their school's internet connection meets their teaching needs. And our teachers do not get enough training and support to integrate technology in their classroom and lessons, despite the fundamental and increasing importance of those skills.

According to the State Educational Technology Directors Association, within a few years schools with at least 1,000 students will need at least 1 Gbps to leverage the transformative power of online learning.[52] By comparison, in South Korea, 100 percent of schools already benefit from ultrafast Internet connections.

The limited bandwidth in America's classrooms is insufficient to support promising applications, such as online tutoring systems or graphic-rich educational games proven to increase student engagement and retention. In fact, most schools cannot support even multiple classrooms simultaneously streaming free educational resources, such as Khan Academy mini-lectures or LearnZillion math lessons. In many cases, bandwidth will also be insufficient to support the interactive tests currently planned across the country that align to the new college and career- ready standards that the Administration has promoted through Race to the Top and the No Child Left Behind reforms.

New education opportunities such as Massive Open Online Courses (MOOCs) and synchronous online learning are enabled by fast broadband speeds at universities and at home, and these education resources are changing how millions of Americans engage in postsecondary learning. With additional broadband capacity, America's K-12 students will also be able to benefit from high-speed opportunities.

Through the Recovery Act's BTOP and BIP grants, the Federal Government sought to help bolster this underserved market. BTOP grant projects will connect nearly 10 percent of all K-12 schools nationwide to networks capable of speeds in excess of 100 Mbps, the speed the U.S. Department of Education cites as a baseline to support 21st century digital learning.

In states like North Carolina and West Virginia, BTOP has been a catalyst for getting virtually all schools connected to high-speed broadband – and the impact of getting schools onto a statewide network is tremendous. In North Carolina, over 70 percent of all schools in the state which did not already have high-capacity broadband available are being connected to a statewide network via the Recovery Act – and consolidating buying power has lowered Internet access prices by about 60 percent. In West Virginia's McDowell County, schools are taking advantage of the improved access to broadband in ways that connect rural students to the entire world and provide rich, interactive learning experiences. In addition to connecting schools to high-speed networks, BTOP projects have offered courses to help teachers integrate online content into their curriculum and provided funding to support devices for students and teachers.

CFY (previously Computers for Youth) is using BTOP funding to provide digital literacy training and computers for low-income sixth graders and their families in New York City and Los Angeles. The CFY program focuses on high-poverty schools, offering Saturday workshops to teach students and parents how to use the Internet and find online educational resources that promote learning in subjects such as math and reading. CFY also trains families on its own PowerMyLearning.com platform, which provides free access to activities and games from across the Web that are designed to make learning fun.

The nation has an opportunity over the next few years to build on the success of BTOP to substantially increase the number of schools connected to high-speed broadband and, in turn, to substantially improve teacher training and digital learning in institutions across the country.

The Connected Initiative

That is why on June 6, 2013, President Obama unveiled the ConnectED initiative, a breakthrough initiative to jumpstart learning technology across our nation's K-12 schools. ConnectED will foster a robust ecosystem for digital learning, supported by robust broadband connectivity and wireless networks connecting 99% of America's K-12 students.

The ConnectED initiative will, within five years, connect 99 percent of America's students, through next-generation broadband (at speeds no less than 100Mbps and with a target of 1Gbps) to, and high-speed wireless within, their schools and libraries. The President is calling on the Federal Communications Commission (FCC) to modernize and leverage the existing E-Rate program, and leverage the expertise of the National

Telecommunications and Information Administration (NTIA) to deliver this connectivity to states, districts, and schools.

Rural communities will experience some of the greatest benefits of new education technologies, as ConnectED will help provide new learning opportunities to level the playing field for rural students. The Universal Service Fund has been transformative in the past twenty years providing rural communities with telephone services, and now broadband. ConnectED builds on BTOP's successes, with greater returns for communities finding it difficult to attract broadband investment.

Innovation in Use of Wireless Spectrum

With demand for spectrum continuing to grow among both commercial and government stakeholders, President Obama remains committed to developing innovative approaches to freeing up more spectrum for wireless broadband. As recognized by the President's Council of Advisers on Science and Technology (PCAST) in its July 2012 report, *Realizing the Full Potential of Government-Held Spectrum to Spur Economic Growth*, commercial and government stakeholders must pursue and invest in emerging technologies and practices that will advance spectrum sharing and other efficiency-enhancing approaches to spectrum policy.[53]

In some cases, Federal systems can be cleared out of a high-value spectrum band and relocated to alternative bands, allowing exclusive commercial access to the valuable band. This strategy is optimal from the perspective of the commercial service and should be pursued where feasible. But in view of the finite nature of spectrum, it is increasingly clear that the nation must augment clearing opportunities with research, development, and deployment of means to promote more spectrum sharing – part of an "all-of-the-above" approach.

Spectrum sharing can take several forms, but all involve two or more users having the right to use the same spectrum band according to rules designed to avoid interference between users. Sharing by Federal agencies maximizes spectrum efficiency and allows faster commercial access while avoiding relocation costs. Some types of sharing are already in common use, while other potentially more promising approaches are being developed and tested. They include:

- **Temporal sharing,** when one user is authorized to use a band of spectrum in some fixed geographic area at certain times and another user has access to the same band at other times.
- **Geographic sharing,** where the government designates particular protection zones restricting commercial access to a particular spectrum band in certain areas, e.g. in the proximity of a military base or a federal satellite receiver.
- **Geo-location/database-enabled sharing,** by which wireless broadband devices use available "white spaces," or gaps, in bands and locations that are allocated to other uses, but only as directed by specific location-aware technologies and databases.
- **Dynamic sensing,** which is emerging as a commercially viable approach as it receives growing attention and investment, and allows networks and devices to react in real time, detecting and moving among a wide range of available spectrum bands in accordance with agreed-upon protocols and priorities.

The continued development and deployment of spectrum sharing, in any of its forms, requires enhanced collaboration between commercial and government users, in the manner that NTIA has been overseeing for the last few years. Sharing of data regarding sensitive government systems presents risks for federal agencies, and they are to be applauded for their efforts to avoid undue risks while still moving in the direction of greater transparency as well as joint spectrum monitoring and testing with the commercial sector.

To accelerate these efforts, the Administration has released a new Presidential Memorandum, *Expanding America's Leadership in Wireless Innovation,* which promotes efficiency, transparency, and collaboration. Building on the recommendations made last year by the PCAST, the memorandum requires Federal agencies to justify system acquisitions and requests for new spectrum assignments in terms of spectrum efficiency and to improve the accuracy of their reporting on spectrum usage. It also directs agencies to increase their collaboration and data-sharing with the private sector and opens up Federal facilities to greater public-private research and development. A newly-formed White House Spectrum Policy Team will oversee implementation of the memorandum and make further recommendations as to spectrum efficiency. In addition, the Administration continues to allocate Federal funding to research and development in spectrum sharing and other advanced communications technologies. Efforts such as these will unleash even greater innovation by our nation's inventors and entrepreneurs as America continues to lead the way in wireless technologies.

End Notes

[1] National Telecommunications and Information Administration. "U.S. Broadband Availability: June 2010 – June 2012." May 2013. Available at http://www.ntia.doc.gov/report/2013/us-broadband-availability-june-2010-june-2012. (Hereafter, NTIA, May 2013).

[2] E.g., the Federal Communications Commission has defined broadband for certain regulatory purposes to mean downstream speeds of at least 4 Mbps and upstream speeds of at least 1 Mbps. Sixth Broadband Report, 25 FCC Rcd 9556, 9563 ¶10 (2010).

[3] Since these lower baselines were used in developing most broadband metrics to date, when determining the "cutoff" for broadband, data throughout this report reflects those earlier metrics. While we believe 10 Mbps downstream is an increasingly "basic" speed, setting a 10 Mbps baseline for future evaluations does not imply that such speeds will fully meet all or even most Americans' needs. As upward evolution in broadband speeds continues, however, reaching that figure at affordable prices should be a low-end standard.

[4] See, for example, U.S. Small Business Administration, Office of Advocacy. "The Impact of Broadband Speed and Price on Small Business." Columbia Telecommunications Corporation, November 2010. Available at: http://archive.sba.gov/advo/research/rs373tot.pdf.

[5] National Telecommunications and Information Administration & Economics and Statistics Administration, "Exploring the Digital Nation: America's Emerging Online Experience." June 2013. Available at http://www.ntia.doc.gov/report/2013/exploring-digital-nation-americas-emerging-online-experience. (Hereafter, NTIA & ESA, June 2013).

[6] Federal Communications Commission, "Eighth Broadband Progress Report.," (August 2012). Available at:http://hraunfoss.fcc.gov/edocs_public/attachmatch/FCC-12-90A1.pdf.

[7] *Open Internet Order*, 25 FCC Rcd 17905 ¶32 (2010).

[8] National Telecommunications and Information Administration, "Digital Nation, Expanding Internet Access, NTIA Research Preview," February 2011. Available at http://www.ntia.doc.gov/files/ntia/publications/ntia_Internet_use_report_february _2011.pdf

[9] Progressive Policy Institute, "Investment Heroes." July 2012. Available at: http://progressive policy.org/wp-content/uploads/2012/07/07.2012-Mandel_Carew_Investment-Heroes_Whos -Betting-on-Americas-Future.pdf.

[10] FCC, "Significant actions and key developments in the Broadband Economy," 2013. Available at: http://webcache. googleusercontent.com/ search?q=cache:Arky8AO3ehwJ: transition. fcc.gov/Daily_Releases/Daily_Business/2013/db0322/DOC-319728A1.doc+&cd=2&hl=en &ct=clnk&gl=us. Deployment of fiber optic cable includes of combination of middle-mile and last mile coverage, thus the increase in new infrastructure may not align with the increase in the availability of broadband via fiber to the premises.

[11] *Ibid.*

[12] NTIA May 2013.

[13] NTIA, May 2013.

[14] LTE and HSPA+ are two types of Terrestrial Mobile Wireless technologies that offer faster speeds than previous generations and are typically considered "4G" in marketing materials. For more information about technology types, see http://www.broadbandmap.gov/classroom/technology.

[15] IDATE Research. "LTE 2013: Markets & Trends, Facts & Figures."

[16] Cisco, Global Mobile Data Traffic Forecast Update, 2012-2017. We note that this report does not compare countries on a country-by-country basis. Akamai's State of the Internet report (below) includes mobile speed data by provider, but not by country. Page 31 of that report shows that of the three mobile providers studied, the average mobile data connection speeds were 1.4 Mbps, 2.6 Mbps and 4.6 Mbps. The fastest mobile provider with the fastest speeds in the world is located in Austria and provides average data connection speeds of 8 Mbps.

[17] Akamai, The State of the Internet, 4th Quarter, 2012 Report, http://www.akamai.com/stateoftheInternet/.

[18] NPD Group, "Internet Connected Devices Surpass Half a Billion in U.S. Homes, According to the NPD Group." March 2013. Available at: https://www.npd.com/wps/portal/ npd/us/news/press-releases/Internet-connected-devices-surpass-half-a-billion-in-u-s-homes-according-to-the-npd-group/.

[19] Aaron Baar, Market Daily. "Tablets, Smartphones Driving CE Sales," July 2012. Available at: http://www.mediapost.com/publications/article/179415/tablets-smartphones-driving-ce-sales.html# axzz2UeuoHIJv. Also FCC, "Significant actions and key developments in the Broadband Economy."

[20] CDC Wireless Substitution. Available at:http://www.cdc.gov/nchs/data/nhis/earlyrelease/wireless201212.pdf.

[21] TechNet, February 2012. Available at: http://www.technet.org/new-technet-sponsored-study-nearly-500000- app-economy-jobs-in-united-states-february-7-2012/.

[22] Austan Goolsbee and Peter Klenow, "Valuing Consumer Products by The Time Spent Using Them: An Application to the Internet," National Bureau of Economic Research, January, 2006.

[23] Yan Chen, Grace YoungJoo Jeon, Yong-Mi Kim, "A Day without a Search Engine: An Experimental Study of Online and Offline Searches," University of Michigan, March 6, 2013.

[24] Erik Brynjolfsson and Joo Hee Oh, "The Attention Economy: Measuring the Value of Free Goods on the Internet."

[25] "The Impact of Broadband on the Economy." International Telecommunication Union, April 2012.

[26] Ericsson, Arthur D. Little, and Chalmers University of Technology, 2011.

[27] Aaron Baar, Market Daily, July 2012. Available at:http://www.mediapost.com/publications/article/179415/tablets-smartphones-driving-ce- sales.html#axzz2UeuoHIJv.

[28] OECD Broadband Portal, Table 1d(2). June 2011. http://www.oecd.org/dataoecd/21/35/39574709.xls.

[29] NTIA & ESA June 2013.

[30] NTIA & ESA June 2013.

[31] NTIA & ESA June 2013.

[32] Rodrigo Belo, "Broadband in Schools: Effects on Student Performance and Spillovers for Household Internet Adoption." 2012. *Dissertations.* Paper 105. Available at: http://repository.cmu.edu/dissertations/105.

[33] FCC, "International Broadband Data Report." Available at: http://www.fcc.gov/document/international- broadband-data-report.

[34] New America Foundation, "The Cost of Connectivity." July 2012. Available at: http://oti.newamerica. net/publications/policy/ the_cost_of_connectivity.

[35] This graphic, available at http://ctcnet.us/DataSpeeds.jpg, provides a visual depiction of the speed capabilities of various types of broadband technology.

[36] NTIA May 2013.

[37] The U.S. Census Bureau categorizes each census block in the country as either "rural" or "urban." The criteria defining the 2010 Census urban areas are available at http://www.census.gov/geo/www/ua/fedregv76n164.pdf; see also http://www.census.gov/geo/www/ua/uafaq.html).

[38] The District of Columbia is not included in these figures because all of its population is urban.

[39] NTIA May 2013.

[40] NTIA May 2013.

[41] For a visualization of the BTOP investments, see NTIA's Connecting America's Communities interactive map at http://www2.ntia.doc.gov/BTOPmap/.

[42] NTIA's Broadband Adoption Toolkit, May 2013. Available at: http://www.ntia.doc.gov/press-release/2013/ntia- broadband-adoption-toolkit-shares-best-practices-across-us.

[43] Edward Wyatt, "F.C.C. Overhauls a Telephone Subsidy," New York Times, October 27, 2011, http://www.nytimes.com/2011/10/28/business/fcc-overhauls-fund-for-broadband-service.html?_r=0. http://hraunfoss.fcc.gov/edocs_public/attachmatch/FCC-11-161A1.pdf

[44] The While House Blog, "President Obama Signs Small Business Jobs Act - Learn What's In It," Sep. 2010. Available at: http://www.whitehouse.gov/blog/2010/09/27/president-obama-signs-small-business-jobs-act-learn-whats-it

[45] The U.S. Department of Treasury, "The Case of Temporary 100 Percent Expensing; Encouraging Business to Expand Now By Lowering the Cost of Investment," Oct. 2010. Available at http://www.whitehouse.gov/sites/default/files/expensing_report.pdf.

[46] The White House Office of the Press Secretary, "Fact Sheet: The Tax Agreement: A Victory for Middle-Class Families and the Economy," Jan. 2013. Available at: http://www.whitehouse.gov/the-press-office/2013/01/01/fact-sheet-tax-agreement-victory-middle-class-families-and-economy.

[47] Kyle McSlarrow and Gail MacKinnon, letter to the Honorable John Boehner, Dec. 2011. Available at http://waysandmeans.house.gov/uploadedfiles/comcast_time_warner.pdf.

[48] Press statement, "ITTA Commends Congress for Passage of Bonus Depreciation," Sep. 2010. Available at http://www.itta.us/advocacy/2010/BONUS%20DEPRECIATION%20SBJCA%2092310%20207%20(f).pdf.

[49] Council of Economic Advisors, "The Economic Benefits of New Spectrum for Wireless Broadband," Feb. 2012. Available at: http://www.whitehouse.gov/sites/default/files/cea_spectrum_report_2-21-2012.pdf.

[50] Cisco's Visual Networking Index: Forecast and Methodology, May 2013. Available at: http://www.cisco.com/en/US/solutions/collateral/ns341/ns525/ns537/ns705/ns827/white_paper_c11-481360_ns827_Networking_Solutions_ White_Paper.html.

[51] Yochai Benkler, "Open Wireless vs. Licensed Spectrum: Evidence from Market Adoption." Available at: http://www.benkler.org/Open_Wireless_V_Licensed_Spectrum_Market_Adoption_current.pdf, p. 6.

[52] State Educational Technology Directors Association, "The Broadband Imperative: Recommendations to Address K-12 Education Infrastructure Needs," 2012. Available at: http://www.setda.org/c/document_library/get_file? folderId=353&name=DLFE-1515.pdf.

[53] PCAST, *"Realizing the Full Potential of Government-Held Spectrum to Spur Economic Growth,"* July 2012. Available at: http://www.whitehouse.gov/sites/default/files/microsites/ostp/pcast_spectrum_report_final_july_20_2012.pdf.

In: Assessing the Digital Nation
Editor: Darius L. Shea ISBN: 978-1-62948-362-7

Chapter 3

THE NATIONAL BROADBAND PLAN GOALS: WHERE DO WE STAND?*

Lennard G. Kruger

SUMMARY

On March 16, 2010, the Federal Communications Commission (FCC) released *Connecting America: The National Broadband Plan*. The National Broadband Plan (NBP) identified significant gaps in broadband availability and adoption in the United States, and in order to address those gaps and other challenges, the NBP set specific goals to be achieved by the year 2020. Goals were set for next generation broadband service; universal broadband service; mobile wireless broadband innovation and coverage; broadband access of Community Anchor Institutions; a nationwide, wireless, interoperable broadband public safety network; and broadband for tracking energy consumption.

Three years after the rollout of the NBP, available data indicate that there has been progress towards reaching the 2020 goals. The following observations can be made:

* This is an edited, reformatted and augmented version of the Congressional Research Service publication, CRS Report for Congress R43016, dated March 19, 2013.

- the United States is much closer to reaching broadband availability goals than broadband adoption goals, which remain a major challenge;
- the United States is much closer to achieving broadband download speed goals than upload speed goals;
- while the next generation broadband goal of 100 million households with 100 Mbps speeds seems within reach (at least for download speeds), the price remains high—affordability could improve in the future depending on technological advances and consumer demand for ultra-high speed next generation performance;
- recent rollouts of next generation wireless technologies have led the FCC to state that the United States leads the world in mobile innovation; on the other hand, the latest Organisation for Economic Co-operation and Development (OECD) data indicate that the United States remains in the middle of the pack with respect to wireless broadband subscriptions per 100 of the population;
- while broadband data are incomplete for Community Anchor Institutions, available information indicate that the number of CAIs with 1 gigabit connections remains relatively low; and
- two major initiatives—FirstNet and Smart Grid—are currently underway in order to help reach the goals for a public safety wireless network and for broadband monitoring of energy consumption.

Many of the key telecommunications issues that are currently being considered by the 113th Congress are focused on improving broadband deployment and thus are intended to have a positive impact on the nation's progress towards reaching one (or in many cases, several) of the NBP goals. As the 113th Congress considers contentious telecommunications issues such as universal service reform, wireless technology and spectrum policy, and telecommunications regulatory reform, the ongoing progress towards meeting the NBP goals is likely to be part of that debate.

INTRODUCTION

On March 16, 2010, the Federal Communications Commission (FCC) released *Connecting America: The National Broadband Plan.*[1] Mandated by the American Recovery and Reinvestment Act of 2009 (ARRA, P.L. 111-5), the FCC's National Broadband Plan (NBP) is a 360-page document composed of 17 chapters containing over 200 specific recommendations directed to the FCC, the Executive Branch (both to individual agencies and to the Administration as a whole), Congress, and nonfederal and nongovernmental entities.

The ARRA mandated that the NBP should "seek to ensure that all people of the United States have access to broadband capability." Accordingly, the NBP identified significant gaps in broadband availability and adoption in the United States, and in order to address those gaps and other challenges, the NBP set six specific goals to be achieved by the year 2020. These goals are as follows:

- Goal No. 1: At least 100 million U.S. homes should have affordable access to actual download speeds of at least 100 megabits per second and actual upload speeds of at least 50 megabits per second.
- Goal No. 2: The United States should lead the world in mobile innovation, with the fastest and most extensive wireless networks of any nation.
- Goal No. 3: Every American should have affordable access to robust broadband service, and the means and skills to subscribe if they so choose.
- Goal No. 4: Every American community should have affordable access to at least 1 gigabit per second broadband service to anchor institutions such as schools, hospitals, and government buildings.
- Goal No. 5: To ensure the safety of the American people, every first responder should have access to a nationwide, wireless, interoperable broadband public safety network.
- Goal No. 6: To ensure that America leads in the clean energy economy, every American should be able to use broadband to track and manage their real- time energy consumption.

This report does not address the appropriateness of those goals or the ongoing debate over the best ways to reach those goals.[2] Rather, this report—with three years having passed since release of the NBP—looks at each of these goals and examines available data and activities which might indicate the nation's progress towards meeting them. Currently, the 113th Congress and the FCC are initiating, developing, and/or overseeing a number of telecommunications policies and programs including universal service reform and a number of spectrum and wireless policy initiatives. Given that those policies and programs are intended to help the nation reach many of the goals set by the NBP, the extent to which the NBP goals are met will likely be a part of the ongoing debate over many of these issues.

GOAL 1. NEXT GENERATION BROADBAND

Goal No. 1: At least 100 million U.S. homes should have affordable access to actual download speeds of at least 100 megabits per second and actual upload speeds of at least 50 megabits per second.

Popularly referred to as "100 squared" (100 Mbps in 100 million households), this goal would offer next-generation broadband[3] to about 75% of all U.S. households. Typically, the higher the speeds offered, the more costly the deployment and the higher the prices to consumers. Deployments of 100 Mbps are more likely in densely populated urban and suburban areas than in rural areas. Currently, capabilities in the range of 100 Mbps/50 Mbps are offered by fiber and a next generation cable technology called DOCSIS 3.0.

According to the most recent National Broadband Map data (released February 2013, data current as of June 2012), **63 million households** are in areas with advertised download speeds greater than 100 Mbps.[4] The same data show that **13 million households** have advertised upload speeds greater than 50 Mbps.[5] The goal also calls for "affordable access." According to the New America Foundation, "it costs between $100 and $300 per month to get a connection speed of over 100 Mbps in most U.S. cities—if such high speeds are even available at all."[6] Similarly, the Information Technology & Information Foundation (ITIF) cites FCC international data indicating an average price in the U.S. of $199.99 per month for 100 Mbps or more. According to ITIF, the high price "reflects the fact that the service tier above 100 Mbps is more a curiosity than a meaningful service today."[7]

It is important to note that the goal sets a target for *actual* speed (e.g., the download and upload speeds actually experienced by consumers), whereas the National Broadband Map data reflects *advertised* speed. The FCC has initiated the Measuring Broadband America Program to survey actual performance data of Internet Service Providers serving over 80% of the residential broadband market. The February 2013 Measuring Broadband America Report found that ISPs deliver on average 97% of advertised speeds during peak intervals.[8]

Meanwhile, the National Broadband Plan set—as a milestone—an interim goal of 100 million homes with actual download speeds of 50 Mbps and actual upload speeds of 20 Mbps by 2015. Accordingly, National Broadband Map data show **101 million households** with advertised download speeds over 50 Mbps, and **22 million households** with advertised upload speeds over 25 Mbps (which is the closest speed tier to 20 Mbps). Thus, while the interim goal has been met for download speeds (100 million households with 50 Mbps by 2015), the interim upload speed goal (100 million households with 20 Mbps) is farther from being reached.

Recently, the cable industry has begun rolling out the latest generation of cable modem technology —DOCSIS 3.0—which can offer 100 Mbps download speeds. Thus, the number of households currently having access to 100 Mbps/50 Mbps is likely higher than what is reflected in the National Broadband Map data, which is current through June 2012. With respect to download speeds, the FCC is optimistic that the 100 squared goal will be met. According to the *Measuring Broadband America* report,

> In just the year since we collected data for our last Report, ISPs have improved in both their ability to deliver what they promise to their customers, and in the overall speeds they can and are delivering. This is a success story, and indicates strong progress toward the important goals set forth in the NBP, that by 2015, 100 million homes should have affordable access to actual download speeds of 50 Mbps, and by 2020 the actual download speed should have increased to 100 Mbps. Though we are making progress toward these goals, we have not yet reached them, and to ensure success it is essential that ISPs continue to improve at the impressive pace indicated by this Report.[9]
>
> Over the next year, we anticipate that providers will continue to innovate and increase their offerings in the higher speed tiers. We know based on industry discussions that the major expansion in high speed service tiers first noted in the July 2012 report was enabled by the deployment by the cable industry of DOCSIS 3.0 technology which permitted service rates of 100 Mbps and above. The cable industry has also announced that it intends in the near future to extend its services to rates beyond 100 Mbps, both to support future service offerings such as ultra-high definition television, and to meet competition created by fiber-based service providers. Verizon fiber is now offering rates up to 300 Mbps in select parts of their market footprint, while Google offers 1 Gbps (1000 Mbps) service in Kansas City, MO.[10]

Similarly, the FCC, in its *Eighth Broadband Progress Report*, notes the acceleration of private sector rollouts of next generation broadband technologies:

> Higher-speed broadband (10 Mbps and above) is increasingly available in many areas of the country. We must keep in mind these developments as we assess the current market and project consumer demand and expectations in the future. For example, cable providers have made much progress on rolling out DOCSIS 3.0, which is capable of 100 Mbps speeds and even higher speeds. And, Americans continue to demand and subscribe to higher services. We will examine in the next Inquiry whether we should identify multiple speed tiers in these reports to assess the country's progress toward our universalization goal, as well as additional goals—such as affordable access to 100 Mbps/50 Mbps to 100 million homes by 2020.[11]

GOAL 2. MOBILE INNOVATION AND WIRELESS NETWORKS

Goal No. 2: The United States should lead the world in mobile innovation, with the fastest and most extensive wireless networks of any nation.

National Broadband Map data monitor the availability and extensiveness of wireless broadband networks in the United States. As **Table 1** shows, lower download speeds (up to 3 Mbps, typically provided by 3G wireless technology) are available to 90% or more of U.S. households. The availability percentage for next generation wireless (4G LTE wireless technologies, which can provide over 6 Mbps) is less, at over 78% of U.S. households. These percentages, reflecting June 2012 data, can be expected to rise, given the recent and ongoing buildout and deployment of 4G LTE wireless broadband technologies.[12]

An assessment of whether or not the United States "lead[s] the world in mobile innovation" is a subjective judgment. The FCC notes that since 2010, U.S. wireless providers have aggressively built out the newest commercial technology for mobile broadband, known as 4G LTE, which offers download speeds in the range of 5 to 12 Mbps. According to the FCC, as of the summer of 2010 there was no LTE deployment in the United States; by January 2012, three mobile wireless providers had launched LTE networks which covered an estimated 211 million people.[13]

The FCC's most recent *International Broadband Data Report* asserts that this recent wireless network building is "securing the United States' position as the world leader in LTE adoption."[14]

Table 1. Nationwide Availability of Mobile Wireless Broadband

Download speeds	Percentage of Household Units
>768 kilobits per second	98.51%
>1.5 megabits per second	95.22%
>3 megabits per second	90.49%
>6 megabits per second	78.62%
>10 megabits per second	76.77%
>25 megabits per second	0.0%

Source: Compiled from National Broadband Map, *Broadband Statistics Report*, "Access to Broadband Technology by Speed," data as of June 2012, released January 2013, available at http://www.broadbandmap.gov/download/Technology %20by%20Speed.pdf.

The FCC notes that according to Deloitte, U.S. investment in 4G networks during 2012-2016 could be $25-$53 billion, that more than 80% of smartphones sold globally run on U.S. operating systems (up from less than 25% three years ago), and that as the first adopters of 4G LTE, the United States is the global test bed for wireless technology and services. In view of this recent progress, the FCC concludes that "the United States has regained its role as a global leader in and around mobile broadband."[15]

On the other hand, another data source, from the Organisation for Economic Co-operation and Development (OECD), provides international comparisons with respect to wireless broadband subscriptions per 100 inhabitants. These data refer to actual subscriptions as opposed to the coverage or extensiveness of wireless networks. The latest OECD data, current as of June 2012, shows the United States ranking 8th behind Korea, Sweden, Australia, Finland, Denmark, Japan, and Norway in wireless broadband subscriptions per 100 inhabitants.[16]

GOAL 3. UNIVERSAL BROADBAND

Goal No. 3: Every American should have affordable access to robust broadband service, and the means and skills to subscribe if they so choose.

This goal encompasses two separate but interrelated and measurable aspects of broadband deployment: availability and adoption. The download and upload speeds at which broadband might be considered "robust" is a subjective judgment that depends on what speeds and associated applications are considered required by broadband users. Section 706(d)(1) of the Telecommunications Act of 1996 (P.L. 104-104) defines "advanced telecommunications capability" as "high-speed, switched, broadband telecommunications capability that enables users to originate and receive high-quality voice, data, graphics, and video telecommunications using any technology."

The National Broadband Plan recommended a broadband benchmark speed of 4 Mbps download/1 Mbps upload, which could initially serve as a minimum threshold constituting broadband under section 706. According to the *Eighth Broadband Progress Report*:

> In each of the reports the Commission has conducted under section 706, it has relied on a speed benchmark for determining whether a service satisfies this statutory definition. In the *2010 Sixth Broadband Progress Report*, the Commission updated this speed benchmark from 200 kbps in both directions to services that offer actual download (i.e., to the customer) speeds of at least 4 Mbps and actual upload (i.e., from the customer) speeds of at least 1 Mbps (4 Mbps/1 Mbps, or "speed benchmark").
>
> In this report, we continue to rely upon this speed benchmark, which the Commission has used in the two most recent broadband reports. We find that this speed benchmark still reflects the requirements in section 706(d)(1) and generally "enables users to originate and receive high-quality voice, data, graphics, and video telecommunications using any technology." For instance, broadband service offering 4 Mbps/1 Mbps enables users to stream high-definition video and engage in basic video conferencing. Maintaining the speed benchmark from prior years also simplifies the measurement of progress from the prior two years.[17]

Because the consumer demand for bandwidth and services increases over time, the NBP recommended that the 4 Mbps download/1 Mbps upload benchmark be reviewed and possibly reset every four years.

Availability

Broadband availability refers to the presence of broadband service that is offered in a given area. According to the National Broadband Map, **98.2% of the U.S. population** has broadband service available with speeds of at least 3

Mbps download/768 kbps upload. This speed level is the National Broadband Map data measurement tier closest to the FCC's 4 Mbps/1 Mbps benchmark. Therefore, the FCC uses the 3 Mbps/768 kbps data as a surrogate for the broadband benchmark. Under this criterion, broadband availability levels appear to be approaching 100%.

Adoption

Broadband adoption refers to the extent to which Americans actually subscribe to and use broadband. Specifically, the NBP set an adoption goal of "higher than 90%" by 2020.[18]

There are several ways to characterize broadband adoption. The FCC's semiannual report tracking broadband subscribership (also known as the Form 477 report) provides subscribership ratios, which is the number of fixed broadband connections nationwide divided by the number of households. According to the most recent semiannual FCC subscriber data (current as of December 31, 2011, released in February 2013), the subscriber ratio is 40% for fixed residential connections[19] with advertised speeds of at least 3 Mbps down and 768 kbps up. The subscribership ratio for fixed residential connections of at least 200 kbps in any direction (which previously was the FCC's minimum benchmark for broadband) is 68%.[20]

The latest Pew Internet and American Life Project telephone survey (conducted in April 2012) found that 66% of Americans surveyed said they have broadband connections at home. The download and upload speeds of those connections was not specified. The Pew data show that broadband adoption rates have slowed and plateaued in recent years—adoption was at 63% in 2009, as compared to 66% in the 2012 survey.[21]Additionally, the 66% rate in the Pew Survey refers to any speed identified as broadband by the survey respondent. The subscribership rate for the NBP benchmark speed of 4 Mbps/1 Mbps is lower (at 40% for fixed residential connections) as reported by the FCC. Thus progress towards meeting the "higher than 90%" goal of the NBP appears to have slowed.[22]

According to the Pew Survey, certain demographic groups tend to have lower broadband adoption rates. These groups include minorities, low income households, the elderly, adults with less educational attainment, and rural populations. Of those surveyed without Internet, almost half said that the main reason they don't go on line is because they don't believe the Internet is relevant to them. In an April 2009 survey, most respondents without broadband said that a drop in price was most likely to get them to switch to broadband.[23] The lack of a computer in the home is also a major reason why households choose not to subscribe to broadband service.[24]

Affordability

Affordability is an important aspect of Goal 3, which calls for "affordable access to robust broadband service." Affordability is also a key factor affecting broadband adoption rates, given that one of the major reasons why people say they choose not to subscribe to a broadband service is its cost.

The question then becomes: at what price should broadband service be considered "affordable?" While measuring affordability is complex, difficult to quantify, and often subjective, one way that broadband service affordability is often assessed is by comparing broadband service prices domestically with rates in other countries. Comparisons of international broadband data can be interpreted in very different ways. Some assert that Americans pay significantly more for broadband than those in other countries, especially for next-generation, very high speed connections.[25] Others assert that broadband prices in the U.S. are reasonable, and that network performance in the U.S. is better than in all but a handful of nations that have densely populated urban areas and have used government subsidies.[26]

The Broadband Data Improvement Act (P.L. 110-385) requires the FCC to prepare international comparisons of broadband service. The third annual *International Broadband Data Report* surveyed broadband plans in 38 countries, including 213 broadband plans (113 fixed, 100 mobile) in the United States. The FCC found that the cheapest plan among those surveyed was $23 per month with 768 kbps download speed and unlimited data. The most expensive standalone broadband plan was a fiber broadband plan at $199 per month with 150 Mbps of download speed, 35 Mbps of upload speed and unlimited data.[27]

Table 2 shows average prices of fixed residential (wireline) broadband plans in the United States surveyed by the FCC. The FCC found that:

> From the analysis of the data, we have concluded that the United States is in the midprice range of countries, whether we compare by speed tier or price per gigabyte of data, for fixed residential broadband.[28]

Table 2. Average Fixed Residential Broadband Prices in the United States

	Average Monthly Price $(2011 PPP[a])	**Rank among OECD and Other Countries with Developed Broadband Markets[c]**
Average monthly net price of a broadband package	$69.75[b]	32nd out of 38
Average monthly net price of a standalone broadband plan, advertised download speed 1-5 Mbps	$34.93	14th out of 24
Average monthly net price of a standalone broadband plan, advertised download speed 5-15 Mbps	$43.71	21st out of 33
Average monthly net price of a standalone broadband plan, advertised download speed 15-25 Mbps	$56.50	26th out of 32

Source: FCC, *Third Annual International Broadband Data Report.*

a. Purchasing power parity (PPP) is a currency conversion rate that converts to a common currency and equalizes the purchasing power of different currencies.

b. This price is a simple average of all the U.S. plans (standalone and bundled broadband) in the dataset.

c. Ranked from lowest to highest average price.

With respect to mobile broadband, the FCC concluded that "particularly for smartphone plans, the United States is one of the ten least expensive countries in terms of price per gigabyte of data."[29]

GOAL 4. ONE GIGABIT TO ANCHOR INSTITUTIONS

Goal No. 4: Every American community should have affordable access to at least 1 gigabit per second broadband service to anchor institutions such as schools, hospitals, and government buildings.

Based on currently available data, it is difficult if not impossible to determine how many communities have affordable access to at least 1 gigabit per second broadband service to their anchor institutions. The National Broadband Map includes a dataset which contains nearly 300,000 records of community anchor institutions (CAIs) including schools, colleges and universities, libraries, medical/healthcare facilities, public safety institutions, and community centers (both governmental and nongovernmental).[30] For a large percentage of the CAIs in the dataset (54%) it is unknown whether or not they subscribe to broadband service. However, of the CAIs which have reported whether or not they have broadband service (and which have reported the download speeds they are receiving), 7.8% report download speeds greater than 1 gigabit per second.

The most recent FCC report on the status of broadband deployment—the *Eighth Broadband Progress Report*—concluded that elementary and secondary schools may lack a sufficient level of broadband service, and that "it continues to appear that many schools and classrooms are underserved by broadband today."[31] Specifically, the FCC cites a January 2011 survey of the Schools and Libraries Program (E-rate) funded schools and libraries in which 80% of E-rate recipients said their broadband connections did not fully meet their needs, and 78% of recipients said they need additional bandwidth. Based on National Broadband Map data, the FCC found that "providers offer download speeds of at least 25 Mbps to only 63.7 percent of the nation's schools, suggesting that many schools may not have a sufficient level of broadband service."[32]

Other data are available from the American Library Association's Information Policy and Access Center, which found that 31.2% of libraries have connectivity speeds greater than 10 Mbps, and 55% have speeds between 1.5 and 10 Mbps.[33]

Meanwhile, one of the purposes of the broadband stimulus grants awarded by the National Telecommunications and Information Administration (NTIA) is to provide ultra-high-speed broadband connections to CAIs. Specifically, the Comprehensive Community Infrastructure grant program (under the Broadband Technology Opportunity Program or BTOP) has awarded 117 grants, many of which are due to be complete by the end of FY2013. According to NTIA, 12,000 CAIs have been connected, and over 80% of the 1,500 communities served by the projects will receive speeds greater than a gigabit per second.[34]

Goal 5. Broadband Public Safety Network

Goal No. 5: To ensure the safety of the American people, every first responder should have access to a nationwide, wireless, interoperable broadband public safety network.

In its 2010 National Broadband Plan report, the FCC stated that "nearly a decade after 9/11, our first responders still lack a nationwide public safety mobile broadband communications network, even though such a network could improve emergency response and homeland security."[35]

Provisions in the Middle Class Tax Relief and Job Creation Act of 2012 (P.L. 112-96) called for developing, constructing, and operating a nationwide network, called FirstNet, designed to meet public safety communications needs.[36]

Goal 6. Broadband and the Smart Grid

Goal No. 6: To ensure that America leads in the clean energy economy, every American should be able to use broadband to track and manage their real-time energy consumption.

According to the FCC, "broadband and advanced communications infrastructure will play an important role in achieving national goals of energy independence and efficiency."[37] In order for this goal to be realized, there must be universal broadband (see Goal 3, above) and, according to the NBP, "the country will need to modernize the electric grid with broadband and advanced communications."[38]

The FCC's central recommendation for reaching this goal involves integrating broadband into the Smart Grid. Currently, the United States "is undertaking a massive communications and information technology buildout to produce the Smart Grid, which the National Institute of Standards and Technology (NIST) defines as the 'two-way flow of electricity and information to create an automated, widely distributed energy delivery network.'"[39]

Title XIII of the Energy Independence and Security Act of 2007 (P.L. 110-140) set forth the policy of the United States "to support the modernization of the nation's electricity transmission and distribution system to maintain a reliable and secure electricity infrastructure." P.L. 110-140 stipulated initiatives for government programs to undertake in Smart Grid investments, including coordinated research, development, demonstration, and information outreach efforts.[40]

Implementation of NBP Recommendations

The National Broadband Plan contained over 200 specific recommendations intended to help achieve the Plan's goals.[41] The NBP's recommendations were directed to the FCC, to Congress, to the Executive Branch (both to individual agencies and to Administration as a whole), and to nonfederal and nongovernmental entities.

The Benton Foundation maintains a database that tracks the implementation of the NBP recommendations.[42] According to Benton, out of a total 218 recommendations, 17% are completed, 40.4% are in progress, 15.6% are started, and 27.1% are not started.

Many of the key telecommunications issues that are currently being considered by the 113th Congress are focused on improving broadband deployment, and thus are intended to have a positive impact on the nation's progress towards reaching one (or in many cases, several) of the NBP goals. Issues include

- Congressional oversight of the FCC's efforts to establish the Connect America Fund and its Mobility and Remote Areas component funds as part of Universal Service Fund reform;[43]
- Congressional oversight of the FCC's efforts to expand the Lifeline Program by allowing subsidies to low-income Americans to be used for broadband, and to modify the Rural Health Care and Schools and Libraries (E-rate) programs, also part of Universal Service Fund reform;[44]
- Congressional and FCC consideration of spectrum policies intended to make more spectrum available for wireless broadband;[45]
- Congressional oversight of the development, construction, and implementation of FirstNet, a nationwide broadband network designed to meet public safety communications needs;[46]
- Congressional oversight of ARRA broadband grant and loan programs and reauthorization of broadband loan programs in the 2013 farm bill;[47] and
- Congressional consideration of possibly revising the current regulatory framework established by the 1996 Telecommunications Act (and its underlying statute, the Communications Act of 1934) in response to the convergence of telecommunications providers and markets and the transition to an Internet Protocol (IP) based network.

CONCLUDING OBSERVATIONS

Three years after the rollout of the National Broadband Plan, available data indicate that there has been progress towards reaching the 2020 goals. The following observations can be made:

- the United States is much closer to reaching broadband availability goals than broadband adoption goals, which remain a major challenge;
- the United States is much closer to achieving broadband download speed goals than upload speed goals;
- while the 100 squared goal seems well within reach (at least for download speeds), its price remains high—affordability could improve in the future depending on technological advances and consumer demand for 100 Mbps plus speeds;
- recent rollouts of next generation wireless technologies (4G LTE) have led the FCC to state that the United States leads the world in mobile innovation; on the other hand, the latest OECD data indicate that the United States remains in the middle of the pack with respect to wireless broadband subscriptions per 100 population;
- while broadband data are incomplete for Community Anchor Institutions, available information indicate that the number of CAIs with 1 gigabit connections remains relatively low; and
- two major initiatives—FirstNet and Smart Grid—are currently underway in order to help reach goals 5 and 6.

In weighing progress towards reaching the goals of the National Broadband Plan, two important considerations should be taken into account. First, the specific broadband goals developed by the FCC are not necessarily universally agreed upon by all stakeholders. For example, the 100 squared goal has been criticized for endorsing an evolution of broadband deployment that could leave rural areas without the next generation broadband service that urban and suburban areas might enjoy. Second, it is impossible to quantify to what extent progress towards reaching these goals is due to the NBP recommendations versus the natural evolution of the broadband market, independent of any impact of the NBP.

Finally, as the 113th Congress considers contentious telecommunications issues such as universal service reform, wireless technology and spectrum policy, and telecommunications regulatory reform, the ongoing progress towards meeting the NBP goals is likely to be part of that debate.

End Notes

[1] Federal Communications Commission, *Connecting America: The National Broadband Plan*, March 2010, 360 pp., available at http://download.broadband.gov/plan/national-broadband-plan.pdf.

[2] For more information on the policy debates surrounding the major NBP recommendations, see CRS Report R41324, *The National Broadband Plan*, by Lennard G. Kruger et al.

[3] A distinction is often made between "current generation" and "next generation" broadband. "Current generation" typically refers to traditionally deployed cable, DSL, and wireless broadband systems, while "next generation" refers to dramatically faster download and upload speeds offered by fiber and advanced cable technologies, and also potentially by future generations of DSL and wireless technologies. Because next generation broadband offers significantly higher speeds and capacities, it can enable more sophisticated and diverse applications.

[4] Derived from National Telecommunications and Information Administration, *Broadband Statistics Report*, "Access to Broadband Technology by Speed," June 2012, available at http://www.broadbandmap.gov/download/ Technology%20by%20Speed.pdf.

[5] National Telecommunications and Information Administration, *Broadband Statistics Report*, "Broadband Availability in Urban vs. Rural Areas," June 2012, available at http://www. broadbandmap.gov/download/Broadband%20Availability%20in%20Rural%20vs%20Urban %20Areas.pdf.

[6] New America Foundation, *The Cost of Connectivity*, July 2012, p. 9, available at http:// newamerica.net/sites/newamerica.net/files/policydocs/The_Cost_of_Connectivity.pdf.

[7] Richard Bennett, Luke A. Stewart, and Robert D. Atkinson, Information Technology & Innovation Foundation (ITIF), *The Whole Picture: Where America's Broadband Networks Really Stand*, February 2013, p. 51, available at http://www2.itif.org/2013-whole-picture-america-broadband-networks.pdf.

[8] Federal Communications Commission, *2013 Measuring Broadband America February Report: A Report on Consumer Wireline Broadband Performance in the U.S.*, February 2013, p. 9, available at http://transition.fcc.gov/cgb/measuringbroadbandreport/2013/Measuring-Broadband-America-feb-2013.pdf.

[9] Federal Communications Commission, *2012 Measuring Broadband America July Report*, July 2013, p. 49, available at http://transition.fcc.gov/cgb/measuringbroadbandreport/2012/ Measuring-Broadband-America.pdf.

[10] *2013 Measuring Broadband America February Report*, p. 52.

[11] Federal Communications Commission, *Eighth Broadband Progress Report*, FCC 12-90, released August 21, 2012, p. 44, available at http://transition.fcc.gov/Daily_Releases/ Daily_Business/2012/db0827/FCC-12-90A1.pdf.

[12] See, for example: AT&T, *News Release*, "AT&T to Invest $14 Billion to Significantly Expand Wireless and Wireline Broadband Networks, Support Future IP Data Growth and New Services," November 7, 2012, available at http://www.att.com/gen/press-room?pid=23506 &cdvn=news&newsarticleid=35661&mapcode=corporate|consumer.

[13] *Eighth Broadband Progress Report*, p. 6-7.

[14] Federal Communications Commission, *International Broadband Data Report*, Third Report, DA 12-1334, August 21, 2012, p. 2, available at http://hraunfoss.fcc.gov/edocs_public/ attachmatch/DA-12-1334A1.pdf.

[15] Ibid.

[16] OECD Broadband Statistics Update, February 7, 2013, available at http://www.oecd.org/sti/broadband/broadband- statistics-update.htm.

[17] *Eighth Broadband Progress Report*, p. 19-20.

[18] *Connecting America: The National Broadband Plan*, p. 10.

[19] Fixed connections do not include wireless broadband. In the *Eighth Broadband Progress Report*, the FCC explains that it has insufficient information to calculate an adoption rate for mobile services, because many households likely have more than one connected mobile device.

[20] Federal Communications Commission, *Internet Access Services: Status as of December 31, 2011*, February 2013, p. 34-35, available at http://transition.fcc.gov/Daily_Releases/Daily_Business/2013/db0207/DOC-318810A1.pdf.

[21] Joanna Brenner and Lee Rainie, Pew Research Center, *Pew Internet: Broadband*, May 24, 2012, available at http://www.pewinternet.org/Commentary/2012/May/Pew-Internet-Broadband.aspx.

[22] The relatively stalled growth in broadband adoption rates is identified and discussed in: John Horrigan, TechNet, *Broadband Adoption in 2012: Little Movement Since '09 & Stakeholders Can Do More to Spur Adoption*, March 20, 2012, 10 pages; and Richard Bennett, Luke A. Stewart, and Robert D. Atkinson, Information Technology & Innovation Foundation (ITIF), *The Whole Picture: Where America's Broadband Networks Really Stand*, February 2013, 75 pages.

[23] *Pew Internet: Broadband.*

[24] U.S. Department of Commerce, National Telecommunications and Information Administration, *Exploring the Digital Nation: Computer and Internet Use at Home*, November 2011, p. 35, available at http://www.ntia.doc.gov/files/ntia/publications/exploring_the_digital_nation_computer_and_internet_use_at_home_11092011.pdf.

[25] See New America Foundation, *The Cost of Connectivity: A comparison of high-speed Internet prices in 22 cities worldwide*, July 2012, 54 pp., available at http://newamerica.net/sites/newamerica.net/files/policydocs/ The_Cost_of_Connectivity.pdf.

[26] *The Whole Picture: Where America's Broadband Networks Really Stand*, p.4.

[27] Federal Communications Commission, *International Broadband Data Report*, Third Report, DA 12-1334, August 21, 2012, p. 5, available at http://hraunfoss.fcc.gov/edocs_public/attachmatch/DA-12-1334A1.pdf.

[28] Ibid, p. 22.

[29] Ibid.

[30] Available at http://www.broadbandmap.gov/data-download.

[31] *Eighth Broadband Progress Report*, p. 59.

[32] Ibid.

[33] Information and Policy Access Center, *Public Libraries & Broadband*, May 31, 2012, available at http://www.plinternetsurvey.org/sites/default/images/Briefs/BroadbandBrief2012May312012-1.pdf.

[34] Testimony of Lawrence Strickling, Assistant Secretary for Communications and Information, National Telecommunications and Information Administration, Hearing before the House Committee on Energy and Commerce, Subcommittee on Communications and Technology, "Is the Broadband Stimulus Working?" February 27, 2013, available at http://www.ntia.doc.gov/speechtestimony/2013/testimony-assistant-secretary-strickling-broadband-stimulus-working.

[35] *Connecting America: The National Broadband Plan*, p. XI.

[36] For more information, see CRS Report R42543, *The First Responder Network and Next-Generation Communications for Public Safety: Issues for Congress*, by Linda K. Moore.

[37] *Connecting America: The National Broadband Plan*, p. 265.

[38] Ibid, p. 27.

[39] Ibid, p. 267.

[40] See http://www.smartgrid.gov. Also see Executive Office of the President, National Science and Technology Council, A Policy Framework for the 21st Century Grid: A Progress Report, February 2013, 11 pp., available at http://www.whitehouse.gov/sites/default/files/microsites/ostp/2013_nstc_grid.pdf and CRS Report R41886, *The Smart Grid and Cybersecurity—Regulatory Policy and Issues*, by Richard J. Campbell.

[41] For an outline of NBP recommendations and NBP recommendations to Congress, See Tables 1 and 2 in CRS Report R41324, *The National Broadband Plan*, by Lennard G. Kruger et al.

[42] The Benton Foundation broadband tracker is available at http://benton.org/initiatives/national_broadband_plan.

[43] For more information see CRS Report R42524, *Rural Broadband: The Roles of the Rural Utilities Service and the Universal Service Fund*, by Angele A. Gilroy and Lennard G. Kruger.

[44] For more information see CRS Report R42846, *Lifeline Telephone Program: Frequently Asked Questions*, by Angele A. Gilroy and Mark Gurevitz and CRS Report R42524, *Rural Broadband: The Roles of the Rural Utilities Service and the Universal Service Fund*, by Angele A. Gilroy and Lennard G. Kruger.

[45] For more information see CRS Report R40674, *Spectrum Policy in the Age of Broadband: Issues for Congress*, by Linda K. Moore.

[46] For more information see CRS Report R42543, *The First Responder Network and Next-Generation Communications for Public Safety: Issues for Congress*, by Linda K. Moore.

[47] For more information see CRS Report RL33816, *Broadband Loan and Grant Programs in the USDA's Rural Utilities Service*, by Lennard G. Kruger and CRS Report R41775, *Background and Issues for Congressional Oversight of ARRA Broadband Awards*, by Lennard G. Kruger.

In: Assessing the Digital Nation
Editor: Darius L. Shea

ISBN: 978-1-62948-362-7

Chapter 4

BROADBAND INTERNET ACCESS AND THE DIGITAL DIVIDE: FEDERAL ASSISTANCE PROGRAMS*

Lennard G. Kruger and Angele A. Gilroy

SUMMARY

The "digital divide" is a term that has been used to characterize a gap between "information haves and have-nots," or in other words, between those Americans who use or have access to telecommunications and information technologies and those who do not. One important subset of the digital divide debate concerns high-speed Internet access and advanced telecommunications services, also known as broadband. Broadband is provided by a series of technologies (e.g., cable, telephone wire, fiber, satellite, wireless) that give users the ability to send and receive data at volumes and speeds far greater than traditional "dial-up" Internet access over telephone lines.

Broadband technologies are currently being deployed primarily by the private sector throughout the United States. While the numbers of new broadband subscribers continue to grow, studies and data suggest that the rate of broadband deployment in urban/suburban and high income areas is outpacing deployment in rural and low-income areas. Some

* This is an edited, reformatted and augmented version of a Congressional Research Service publication, CRS Report for Congress RL30719, prepared for Members and Committees of Congress, from www.crs.gov, dated July 17, 2013.

policymakers, believing that disparities in broadband access across American society could have adverse economic and social consequences on those left behind, assert that the federal government should play a more active role to avoid a "digital divide" in broadband access.

With the conclusion of the grant and loan awards established by the American Recovery and Reinvestment Act of 2009 (P.L. 111-5), there remain two ongoing federal vehicles which direct federal money to fund broadband infrastructure: the broadband and telecommunications programs at the Rural Utilities Service (RUS) of the U.S. Department of Agriculture and the Universal Service Fund (USF) programs under the Federal Communications Commission (FCC). Although the USF's High Cost Program does not explicitly fund broadband infrastructure, subsidies are used, in many cases, to upgrade existing telephone networks so that they are capable of delivering high-speed services. Additionally, subsidies provided by USF's Schools and Libraries Program and Rural Health Care Program are used for a variety of telecommunications services, including broadband access. Currently the USF is undergoing a major transition to the Connect America Fund, which is targeted to the deployment, adoption, and utilization of both fixed and mobile broadband.

To the extent that the 113th Congress may consider various options for further encouraging broadband deployment and adoption, a key issue is how to strike a balance between providing federal assistance for unserved and underserved areas where the private sector may not be providing acceptable levels of broadband service, while at the same time minimizing any deleterious effects that government intervention in the marketplace may have on competition and private sector investment.

INTRODUCTION

The "digital divide" is a term used to describe a perceived gap between "information haves and have-nots," or in other words, between those Americans who use or have access to telecommunications and information technologies and those who do not.[1] Whether or not individuals or communities fall into the "information haves" category depends on a number of factors, ranging from the presence of computers in the home, to training and education, to the availability of affordable Internet access.

Broadband technologies are currently being deployed primarily by the private sector throughout the United States. While the numbers of new broadband subscribers continue to grow, studies and data suggest that the rate of broadband deployment in urban/suburban and high income areas is outpacing deployment in rural and low-income areas.

STATUS OF BROADBAND DEPLOYMENT IN THE UNITED STATES

Prior to the late 1990s, American homes accessed the Internet at maximum speeds of 56 kilobits per second by dialing up an Internet Service Provider (such as AOL) over the same copper telephone line used for traditional voice service. A relatively small number of businesses and institutions used broadband or high speed connections through the installation of special "dedicated lines" typically provided by their local telephone company. Starting in the late 1990s, cable television companies began offering cable modem broadband service to homes and businesses. This was accompanied by telephone companies beginning to offer DSL service (broadband over existing copper telephone wireline). Growth has been steep, rising from 2.8 million high speed lines reported as of December 1999, to 243 million connections as of June 30, 2012.[2] Of the 243 million high speed connections reported by the FCC, 197 million serve residential users.[3]

Table 1 depicts the relative deployment of different types of broadband technologies. A distinction is often made between "current generation" and "next generation" broadband (commonly referred to as next generation networks or NGN). "Current generation" typically refers to currently deployed cable, DSL, and many wireless systems, while "next generation" refers to dramatically faster download and upload speeds offered by fiber technologies and also by successive generations of cable, DSL, and wireless technologies.[4] In general, the greater the download and upload speeds offered by a broadband connection, the more sophisticated (and potentially valuable) the application that is enabled.

Based on the latest FCC broadband connection data, *Table 2* shows the percentages of households with broadband connections by state, both for download connections over 200 kbps and for connections of at least 3 Mbps (which approximates the FCC's National Broadband Availability target). According to the FCC, high speed connections over 200 kbps are reported in 67% of households nationwide, while connections of at least 3 Mbps (download) and 768 kbps (upload) are reported in 38% of households nationwide.

Table 1. Percentage of Broadband Technologies by Types of Connection

	Connections over 200 kbps in at least one direction	Residential connections over 200 kbps in at least one direction	Connections at least 3 Mbps downstream and 768 kbps upstream	Residential connections at least 3 Mbps downstream and 768 kbps upstream
cable modem	20.4%	23.9%	37.7%	45.3%
DSL	12.8%	14.1%	12.8%	14.7%
Mobile wireless	63.0%	58.2%	42.9%	32.5%
Fiber	2.6%	3.0%	6.0%	7.2%
All other	1.2%	0.8%	0.6%	0.4%

Source: FCC, Internet Access Services: Status as of June 30, 2012, pp. 23-26.

Table 2. Percentage of Households With Broadband Connections by State (as of June 30, 2012)

	Connections over 200 kbps	Connections at least 3 mbps downstream and 768 kbps upstream
Alabama	57%	30%
Alaska	71%	34%
Arizona	69%	54%
Arkansas	54%	25%
California	76%	44%
Colorado	76%	61%
Connecticut	79%	58%
Delaware	78%	69%
District of Columbia	72%	65%
Florida	74%	51%
Georgia	65%	42%
Hawaii	*	*
Idaho	61%	30%
Illinois	67%	44%
Indiana	64%	41%
Iowa	66%	25%
Kansas	67%	33%
Kentucky	62%	38%
Louisiana	60%	30%
Maine	75%	26%
Maryland	76%	70%
Massachusetts	81%	75%
Michigan	65%	46%

	Connections over 200 kbps	Connections at least 3 mbps downstream and 768 kbps upstream
Minnesota	69%	48%
Mississippi	50%	21%
Missouri	62%	29%
Montana	64%	38%
Nebraska	68%	45%
Nevada	66%	50%
New Hampshire	81%	65%
New Jersey	82%	76%
New Mexico	59%	37%
New York	76%	54%
North Carolina	69%	17%
North Dakota	70%	41%
Ohio	68%	25%
Oklahoma	58%	28%
Oregon	69%	55%
Pennsylvania	71%	58%
Rhode Island	76%	*
South Carolina	64%	29%
South Dakota	67%	47%
Tennessee	58%	39%
Texas	65%	33%
Utah	73%	56%
Vermont	77%	61%
Virginia	70%	60%
Washington	72%	61%
West Virginia	60%	49%
Wisconsin	67%	30%
Wyoming	66%	44%
National subscribership ratio	69%	45%

Source: FCC, Internet Access Services: Status as of June 30, 2012, pp. 34-35.

Notes: Asterisk (*) indicates data withheld by the FCC to maintain firm confidentiality. Subscribership ratio is the number of reported residential high speed lines (broadband connections) divided by the number of households in each state.

Meanwhile, the National Broadband Map, which is composed of state broadband data and compiled by NTIA, provides data on where broadband is and is not available. The latest update of these data indicate that 98.2% of the U.S. population have available advertised speeds of at least 3 Mbps (download) and 768 kbps (upload), 96.2% have download speeds 6 Mbps or greater, and 94.4% have download speeds 10 Mbps or greater.[5] The FCC's *Eighth Broadband Progress Report*, released on August 21, 2012, used National Broadband Map data to estimate that 19 million Americans living in 7 million households lack access to fixed broadband at speeds of 4 Mbps (download)/1 Mbps (upload) or greater.[6] *Table 3* shows a state-by-state breakdown of the percentage of population without access to fixed broadband at the FCC's benchmark speed of 4 Mbps/1Mbps.

Table 3. Americans Without Access to Fixed Broadband by State (access to speeds of at least 4 Mbps download/1 Mbps upload)

	% of population without access	% of population without access, nonrural areas	% of population without access, rural areas
United States	6.0%	1.8%	23.7%
Alabama	11.4%	1.6%	25.5%
Alaska	19.6%	4.4%	48.9%
Arizona	4.7%	1.2%	35.8%
Arkansas	13.6%	1.8%	28.8%
California	3.3%	1.6%	35.2%
Colorado	4.3%	1.0%	25.3%
Connecticut	0.7%	0.5%	2.6%
Delaware	3.1%	1.1%	13.0%
District of Columbia	0.0%	0.0%	N/A
Florida	3.1%	2.0%	14.3%
Georgia	3.4%	1.3%	9.9%
Hawaii	1.5%	0.1%	17.7%
Idaho	13.1%	1.3%	41.4%
Illinois	3.3%	0.4%	25.6%
Indiana	4.3%	1.3%	12.4%
Iowa	7.1%	0.7%	18.7%
Kansas	7.7%	1.0%	27.0%
Kentucky	10.5%	1.5%	23.0%
Louisiana	8.8%	1.3%	29.6%
Maine	4.7%	1.2%	7.0%
Maryland	3.2%	0.9%	19.2%
Massachusetts	1.0%	0.5%	6.4%
Michigan	6.3%	0.8%	22.4%
Minnesota	8.0%	0.8%	27.7%
Mississippi	12.1%	1.2%	22.8%
Missouri	7.5%	0.6%	24.2%
Montana	26.7%	4.0%	55.4%

	% of population without access	% of population without access, nonrural areas	% of population without access, rural areas
Nebraska	10.1%	1.9%	33.0%
Nevada	2.3%	0.6%	30.2%
New Hampshire	7.5%	2.5%	15.2%
New Jersey	0.7%	0.4%	5.6%
New Mexico	14.2%	4.8%	46.7%
New York	1.3%	0.0%	10.4%
North Carolina	6.4%	2.1%	15.0%
North Dakota	15.9%	2.5%	36.2%
Ohio	3.4%	0.5%	14.0%
Oklahoma	16.2%	2.9%	42.5%
Oregon	3.4%	0.2%	17.3%
Pennsylvania	1.7%	0.3%	6.8%
Rhode Island	0.2%	0.0%	2.3%
South Carolina	11.7%	4.9%	25.1%
South Dakota	21.1%	3.2%	44.6%
Tennessee	6.8%	0.9%	18.6%
Texas	5.9%	2.0%	27.6%
Utah	1.8%	0.3%	16.7%
Vermont	9.4%	0.2%	15.2%
Virginia	10.9%	2.2%	37.6%
Washington	3.2%	0.5%	17.4%
West Virginia	45.9%	31.4%	59.8%
Wisconsin	6.9%	0.1%	23.0%
Wyoming	13.2%	1.1%	35.4%
U.S. Territories	54.0%	41.5%	85.2%
American Samoa	78.6%	30.9%	92.0%
Northern Marianas	100.0%	100.0%	100.0%
Guam	54.3%	0.1%	76.1%
Puerto Rico	51.6%	40.3%	84.8%
U.S. Virgin Islands	100.0%	100.0%	100.0%

Source: FCC, Eighth Broadband Progress Report, Appendix C.

In contrast to broadband *availability*, which refers to whether or not broadband service is offered, broadband *adoption* refers to the extent to which American households actually subscribe to and use broadband. The U.S. Department of Commerce, based on October 2012 survey data, found that 72.4% of U.S. households have adopted broadband.[7] Similarly, the FCC's *Eighth Broadband Progress Report* found that 64% of American households with broadband available to them adopt broadband service offering speeds faster than 768 kbps/200 kbps, while 40% adopt speeds faster than the FCC benchmark of 4 Mbps/1Mbps. The FCC found that the "broadband adoption rates for American households are lower, on average, in the counties with the lowest median household income, in areas outside of urban areas, on Tribal lands, and in U.S. Territories."[8]

According to the Department of Commerce report *Exploring the Digital Nation: America's Emerging Online Experience* (based on July 2011 U.S. Census Bureau survey data), the three main reasons cited for not having broadband at home are that it is perceived as not needed, too expensive, and lack of a home computer.[9] The Department of Commerce report, the FCC's National Broadband Plan, and a survey conducted by the Pew Internet and American Life Project[10] also found disparities in broadband adoption among demographic groups. Populations continuing to lag behind in broadband adoption include people with low incomes, seniors, minorities, the less-educated, non-family households, and the non-employed.

BROADBAND IN RURAL AREAS[11]

While the number of new broadband subscribers continues to grow, the rate of broadband deployment in urban areas appears to be outpacing deployment in rural areas.

While there are many examples of rural communities with state of the art telecommunications facilities,[12] recent surveys and studies have indicated that, in general, rural areas tend to lag behind urban and suburban areas in broadband deployment. For example:

- According to the FCC's *Eighth Broadband Progress Report*, released in August 2012, of the 19 million Americans who live where fixed broadband is unavailable, 14.5 million live in rural areas.[13]
- According to June 2012 data from the National Broadband Map, 99.6% of the population in urban areas have access to available broadband download speeds of at least 6 Mbps, as opposed to 81.8% of the population in rural areas.[14]
- The Department of Commerce report, *Exploring the Digital Nation: America's Emerging Online Experience*, found that while the digital divide between urban and rural areas has lessened since 2007, it still

persists with 72% of urban households adopting broadband service in 2011, compared to 58% of rural households.[15]

The comparatively lower population density of rural areas is likely the major reason why broadband is less deployed than in more highly populated suburban and urban areas. Particularly for wireline broadband technologies—such as cable modem and DSL—the greater the geographical distances among customers, the larger the cost to serve those customers. Thus, there is often less incentive for companies to invest in broadband in rural areas than, for example, in an urban area where there is more demand (more customers with perhaps higher incomes) and less cost to wire the market area.[16]

Some policymakers believe that disparities in broadband access across American society could have adverse consequences on those left behind, and that advanced telecommunications applications critical for businesses and consumers to engage in e-commerce are increasingly dependent on high speed broadband connections to the Internet. Thus, some say, communities and individuals without access to broadband could be at risk to the extent that connectivity becomes a critical factor in determining future economic development and prosperity. A February 2006 study done by the Massachusetts Institute of Technology for the Economic Development Administration of the Department of Commerce marked the first attempt to quantitatively measure the impact of broadband on economic growth. The study found that "between 1998 and 2002, communities in which mass-market broadband was available by December 1999 experienced more rapid growth in employment, the number of businesses overall, and businesses in IT-intensive sectors, relative to comparable communities without broadband at that time."[17]

A June 2007 report from the Brookings Institution found that for every one percentage point increase in broadband penetration in a state, employment is projected to increase by 0.2% to 0.3% per year. For the entire U.S. private non-farm economy, the study projected an increase of about 300,000 jobs.[18]

Subsequently, a July 2009 study commissioned by the Internet Innovation Alliance found net consumer benefits of home broadband on the order of $32 billion per year, up from an estimated $20 billion in consumer benefits from home broadband in 2005.[19]

Some also argue that broadband is an important contributor to U.S. future economic strength with respect to the rest of the world. Data from the Organization for Economic Cooperation and Development (OECD) found the U.S. ranking 15th among OECD nations in broadband access per 100 inhabitants as of December 2011.[20] By contrast, in 2001 an OECD study found the U.S. ranking fourth in broadband subscribership per 100 inhabitants (after Korea, Sweden, and Canada).[21] While many argue that declining U.S. performance in international broadband rankings is a cause for concern,[22] others maintain that the OECD data undercount U.S. broadband deployment,[23] and that cross-country broadband deployment comparisons are not necessarily meaningful and are inherently problematic.[24] Finally, an issue related to international broadband rankings is the extent to which broadband speeds and prices differ between the United States and the rest of the world.[25]

ARE BROADBAND DEPLOYMENT DATA ADEQUATE?

Obtaining an accurate snapshot of the status of broadband deployment is problematic. Anecdotes abound of rural and low-income areas which do not have adequate Internet access, as well as those which are receiving access to high-speed, state-of-the-art connections. Rapidly evolving technologies, the constant flux of the telecommunications industry, the uncertainty of consumer wants and needs, and the sheer diversity and size of the nation's economy and geography make the status of broadband deployment very difficult to characterize. The FCC periodically collects broadband deployment data from the private sector via "FCC Form 477"—a standardized information gathering survey. Statistics derived from the Form 477 survey are published every six months. Additionally, data from Form 477 are used in the FCC's (to date) eight broadband deployment reports.

The FCC is working to refine the data used in future reports in order to provide an increasingly accurate portrayal. In its March 17, 2004, Notice of Inquiry for the *Fourth Report*, the FCC sought comments on specific proposals to improve the FCC Form 477 data gathering program.[26] On November 9, 2004, the FCC voted to expand its data collection program by requiring reports from all facilities based carriers regardless of size in order to better track rural and underserved markets, by requiring broadband providers to provide more information on the speed and nature of their service, and by establishing broadband-over-power line as a separate category in order to track its development and deployment. The FCC Form 477 data gathering program was extended for five years beyond its then March 2005 expiration date.[27]

On April 16, 2007, the FCC announced a Notice of Proposed Rulemaking which sought comment on a number of broadband data collection issues, including how to develop a more accurate picture of broadband deployment; gathering information on price, other factors determining consumer uptake of broadband, and international comparisons; how to improve data on wireless broadband; how to collect information on subscribership to voice over Internet Protocol service (VoIP); and whether to modify collection of speed tier information.[28]

On March 19, 2008, the FCC adopted an order that substantially expands its broadband data collection capability. Specifically, the order expands the number of broadband reporting speed tiers to capture more information about upload and download speeds offered in the marketplace, requires broadband providers to report numbers of broadband subscribers by census tract, and improves the accuracy of information collected on mobile wireless broadband deployment.

Additionally, in a Further Notice of Proposed Rulemaking, the FCC sought comment on broadband service pricing and availability.[29] The July 2009 data release (providing data as of June 30, 2008) was the final data set gathered under the old FCC Form 477. The February 2010 data report (December 31, 2008, data) reflected the new Form 477 data collection requirements. Meanwhile, during the 110th Congress, state initiatives to collect broadband deployment data in order to promote broadband in underserved areas were viewed as a possible model for governmental efforts to encourage broadband.

The Broadband Data Improvement Act was enacted by the 110th Congress and became P.L. 110-385 on October 10, 2008. The law requires the FCC to collect demographic information on unserved areas, data comparing broadband service with 75 communities in at least 25 nations abroad, and data on consumer use of broadband. The act also directs the Census Bureau to collect broadband data, the Government Accountability Office to study broadband data metrics and standards, and the Department of Commerce to provide grants supporting state broadband initiatives.

P.L. 111-5, the American Recovery and Reinvestment Act, provided NTIA with an appropriation of $350 million to implement the Broadband Data Improvement Act and to develop and maintain a national broadband inventory map. The National Broadband Map was first released on February 17, 2011 (http://www.broadbandmap.gov), and is updated every six months.[30]

Finally, the FCC's National Broadband Plan addressed the broadband data issue, recommending that the FCC and the U.S. Bureau of Labor Statistics (BLS) should collect more detailed and accurate data on actual availability, penetration, prices, churn, and bundles offered by broadband service providers to consumers and businesses, and should publish analyses of these data.

BROADBAND AND THE FEDERAL ROLE

The Telecommunications Act of 1996 (P.L. 104-104) addressed the issue of whether the federal government should intervene to prevent a "digital divide" in broadband access. Section 706 requires the FCC to determine whether "advanced telecommunications capability [i.e., broadband or high-speed access] is being deployed to all Americans in a reasonable and timely fashion."

Since 1999, the FCC has adopted and released eight reports pursuant to Section 706. The first five reports formally concluded that the deployment of advanced telecommunications capability to all Americans is reasonable and timely. Unlike the first five 706 reports, the sixth, seventh, and eighth reports concluded that broadband is not being deployed to all Americans in a reasonable and timely fashion. According to the *Eighth Broadband Progress Report*:

> Our analysis shows that the nation's broadband deployment gap remains significant and is particularly pronounced for Americans living in rural areas and on Tribal lands. We find that as of June 30, 2011, approximately 19 million Americans did not have access to fixed broadband. Significantly, approximately 76 percent of these Americans reside in rural areas. Our analysis further shows that Americans residing on Tribal lands disproportionately lack access to fixed broadband. And the available international broadband data, though not perfectly comparable to U.S. data, suggest that the availability and deployment of broadband in the United States may lag behind a number of other developed countries in certain respects, although we also compare favorably to some developed countries in other respects. Moreover, as many as 80 percent of E-rate recipients say that their broadband connections do not fully meet their needs, and 78 percent of recipients say that they need additional bandwidth. These data combined with our findings concerning availability above provide further indication that broadband is not yet being reasonably and timely deployed to all Americans.[31]

FCC Commissioners Robert McDowell and Ajit Pai issued dissenting statements, maintaining that there is insufficient justification for the 706 report conclusion that broadband is not being deployed in a reasonable and timely fashion. For example, the dissents argued that the report did not sufficiently account for the dramatic growth in the availability and deployment of mobile broadband, and that gaps in broadband adoption should not be used to determine whether or not broadband is being sufficiently deployed.[32]

The National Broadband Plan

As mandated by the ARRA, on March 16, 2010, the FCC publically released its report, *Connecting America: The National Broadband Plan.*[33] The National Broadband Plan (NBP) seeks to "create a high-performance America," which the FCC defines as "a more productive, creative, efficient America in which affordable broadband is available everywhere and everyone has the means and skills to use valuable broadband applications."[34] In order to achieve this mission, the NBP recommends that the country set six goals for 2020:

- Goal No. 1: At least 100 million U.S. homes should have affordable access to actual download speeds of at least 100 megabits per second and actual upload speeds of at least 50 megabits per second.
- Goal No. 2: The United States should lead the world in mobile innovation, with the fastest and most extensive wireless networks of any nation.
- Goal No. 3: Every American should have affordable access to robust broadband service, and the means and skills to subscribe if they so choose.
- Goal No. 4: Every American community should have affordable access to at least 1 gigabit per second broadband service to anchor institutions such as schools, hospitals, and government buildings.
- Goal No. 5: To ensure the safety of the American people, every first responder should have access to a nationwide, wireless, interoperable broadband public safety network.
- Goal No. 6: To ensure that America leads in the clean energy economy, every American should be able to use broadband to track and manage their real-time energy consumption.

The National Broadband Plan is categorized into three parts:

- *Part I (Innovation and Investment),* which "discusses recommendations to maximize innovation, investment and consumer welfare, primarily through competition. It then recommends more efficient allocation and management of assets government controls or influences."[35] The recommendations address a number of issues, including spectrum policy, improved broadband data collection, broadband performance standards and disclosure, special access rates, interconnection, privacy and cybersecurity, child online safety, poles and rightsof-way, research and experimentation (R&E) tax credits, and R&D funding.

- *Part II (Inclusion),* which "makes recommendations to promote inclusion—to ensure that all Americans have access to the opportunities broadband can provide."[36] Issues include reforming the Universal Service Fund, intercarrier compensation, federal assistance for broadband in Tribal lands, expanding existing broadband grant and loan programs at the Rural Utilities Service, enabling greater broadband connectivity in anchor institutions, and improved broadband adoption and utilization especially among disadvantaged and vulnerable populations.
- *Part III (National Purposes),* which "makes recommendations to maximize the use of broadband to address national priorities. This includes reforming laws, policies and incentives to maximize the benefits of broadband in areas where government plays a significant role."[37] National purposes include health care, education, energy and the environment, government performance, civic engagement, and public safety. Issues include telehealth and health IT, online learning and modernizing educational broadband infrastructure, digital literacy and job training, smart grid and smart buildings, federal support for broadband in small businesses, telework within the federal government, cybersecurity and protection of critical broadband infrastructure, copyright of public digital media, interoperable public safety communications, next generation 911 networks, and emergency alert systems.

The release of the National Broadband Plan is seen by many as a precursor towards the development of a national broadband policy—whether comprehensive or piecemeal—that will likely be shaped and developed by Congress, the FCC, and the Administration.[38] Congress will likely play a major role in implementing the National Broadband Plan, both by considering legislation to implement NBP recommendations, and by overseeing broadband activities conducted by the FCC and executive branch agencies.

CURRENT FEDERAL BROADBAND PROGRAMS

With the conclusion of grant and loan awards established by the American Recovery and Reinvestment Act of 2009 (P.L. 111-5),[39] there remain two ongoing federal vehicles which direct federal money to fund broadband infrastructure: the broadband and telecommunications programs at the Rural Utilities Service (RUS) of the U.S. Department of Agriculture and the Universal Service Fund (USF) programs under the Federal Communications Commission (FCC). Although the USF's High Cost Program does not explicitly fund broadband infrastructure, subsidies are used, in many cases, to upgrade existing telephone networks so that they are capable of delivering high-speed services. Additionally, subsidies provided by USF's Schools and Libraries Program and Rural Health Care Program are used for a variety of telecommunications services, including broadband access. Currently the USF is undergoing a major transition to the Connect America Fund, which is targeted to the deployment, adoption, and use of both fixed and mobile broadband.

Table 4 (at the end of this report) shows selected federal domestic assistance programs throughout the federal government that currently can be associated with broadband and telecommunications development. The table categorizes the programs in three ways: programs exclusively devoted to the deployment of broadband infrastructure; programs which have traditionally focused on deployment of telecommunications infrastructure generally (which typically can and does include broadband); and applications-specific programs which fund some aspect of broadband access or adoption as a means towards supporting a particular application, such as distance learning or telemedicine.

Rural Utilities Service Programs

RUS implements two programs specifically targeted at providing assistance for broadband infrastructure deployment in rural areas: the Rural Broadband Access Loan and Loan Guarantee Program and Community Connect Broadband Grants.[40] The 110th Congress reauthorized and reformed the Rural Broadband Access Loan and Loan Guarantee program as part of the 2008 farm bill (P.L. 110-234). The 112th Congress considered reauthorization of the program as part of the 2012 farm bill.[41]

RUS also has a rural telephone loan program (dating back to 1949, now called Telecommunications Infrastructure Loans) that has historically supported infrastructure for telephone voice service, but has now evolved into support for broadband-capable service provided by traditional telephone borrowers. Additionally, the Distance Learning and Telemedicine Grant Program supports broadband-based applications.[42]

The Universal Service Concept and the FCC[43]

Since its creation in 1934 the Federal Communications Commission (FCC) has been tasked with "mak[ing] available, so far as possible, to all the people of the United States ... a rapid, efficient, Nation-wide, and world-wide wire and radio communications service with adequate facilities at reasonable charges."[44] This mandate led to the development of what has come to be known as the universal service concept.

The universal service concept, as originally designed, called for the establishment of policies to ensure that telecommunications services are available to all Americans, including those in rural, insular and high cost areas, by ensuring that rates remain affordable. Over the years this concept fostered the development of various FCC policies and programs to meet this goal. The FCC offers universal service support through a number of direct mechanisms that target both providers of and subscribers to telecommunications and, more recently, broadband services.[45]

Universal Service and the Telecommunications Act of 1996

Passage of the Telecommunications Act of 1996 (P.L. 104-104) codified the long-standing commitment by U.S. policymakers to ensure universal service in the provision of telecommunications services.

The Schools and Libraries, and Rural Health Care Programs

Congress, through the 1996 act, not only codified, but also expanded the concept of universal service to include, among other principles, that elementary and secondary schools and classrooms, libraries, and rural health care providers have access to telecommunications services for specific purposes at discounted rates. (See §§254(b)(6) and 254(h)of the 1996 Telecommunications Act, 47 U.S.C. 254.)

1. The Schools and Libraries Program. Under universal service provisions contained in the 1996 act, elementary and secondary schools and classrooms and libraries are designated as beneficiaries of universal service discounts. Universal service principles detailed in Section 254(b)(6) state that

"Elementary and secondary schools and classrooms ... and libraries should have access to advanced telecommunications services." The act further requires in Section 254(h)(1)(B) that services within the definition of universal service be provided to elementary and secondary schools and libraries for education purposes at discounts, that is at "rates less than the amounts charged for similar services to other parties."

The FCC established the Schools and Libraries Division within the Universal Service Administrative Company (USAC) to administer the schools and libraries or "E (education)-rate" program to comply with these provisions. Under this program, eligible schools and libraries receive discounts ranging from 20% to 90% for telecommunications services depending on the poverty level of the school's (or school district's) population and its location in a high cost telecommunications area. The FCC established a funding ceiling, or cap, of $2.25 billion, adjusted for inflation prospectively beginning with funding year 2010. Three categories of services are eligible for discounts: internal connections (e.g., wiring, routers and servers); Internet access; and telecommunications and dedicated services, with the third category receiving funding priority. According to data released by program administrators, approximately $31 billion in funding has been committed over the first 14 years of the program with funding released to all states, the District of Columbia and all territories. Funding commitments for funding Year 2012 (July 1, 2012-June 30, 2013), the 15th year of the program, totaled $ 2.6 billion as of July 5, 2013.[46]

2. The Rural Health Care Programs. Section 254(h) of the 1996 act requires that public and nonprofit rural health care providers have access to telecommunications services necessary for the provision of health care services at rates comparable to those paid for similar services in urban areas. Subsection 254(h)(1) further specifies that "to the extent technically feasible and economically reasonable" health care providers should have access to advanced telecommunications and information services. The FCC established the Rural Health Care Division (RHCD) within the USAC to administer the universal support program to comply with these provisions. Under FCC established rules only public or non-profit health care providers are eligible to receive funding. Eligible health care providers, with the exception of those requesting only access to the Internet, must also be located in a rural area. The funding ceiling, or cap, for this support was established at $400 million annually. The funding level for Year One of the program (January 1998-June 30, 1999) was set at $100 million. Due to less than anticipated demand, the FCC established a $12 million funding level for the second year (July 1, 1999

to June 30, 2000) of the program but has since returned to a $400 million yearly cap. As of March 31, 2012, a total of $514.3 million has been committed since the program's inception in 1998. The primary use of the funding is to provide reduced rates for telecommunications and information services necessary for the provision of health care.[47] In addition, the FCC established, in 2007, the "Rural Health Care Pilot Program" to help public and non-profit health care providers build state and region-wide broadband networks dedicated to the provision of health care services. There are 50 projects in the program with $387.9 million in authorized funds. As of February 29, 2010, $232.6 million of the funds have been committed to the 50 FCC designated projects. The FCC in a December 12, 2012, order, created a new program, the Healthcare Connect Fund, which will expand health care provider access to broadband, particularly in rural areas, and replace the Rural Health Care Pilot Program with a permanent program. Non-rural participation is limited to consortia of which at least 50% must be located in a rural area. The upfront funding cap for the Healthcare Connect Fund is $150 million annually and participants will be required to contribute 35% of the costs. The FCC also established, as part of the Healthcare Connect Fund, a new pilot program, to expand broadband connections to skilled nursing facilities. Funding for this pilot program, which will begin in 2014, will be up to $50 million total over three years. The total funding cap for all of the above mentioned USF rural health care programs will remain at $400 million annually.[48]

Universal Service and Broadband

One of the policy debates surrounding universal service in the last decade was whether access to advanced telecommunications services (i.e., broadband) should be incorporated into universal service objectives. The term universal service, when applied to telecommunications, refers to the ability to make available a basket of telecommunications services to the public, across the nation, at a reasonable price.

As directed in the 1996 Telecommunications Act (§254[c]) a federal-state Joint Board was tasked with defining the services which should be included in the basket of services to be eligible for federal universal service support; in effect using and defining the term "universal service" for the first time. The Joint Board's recommendation, which was subsequently adopted by the FCC in May 1997, included the following in its universal service package: voice grade access to and some usage of the public switched network; single line service; dual tone signaling; access to directory assistance; emergency service such as 911; operator services; and access and interexchange (long distance) service.

Some policy makers expressed concern that the FCC-adopted definition is too limited and does not take into consideration the importance and growing acceptance of advanced services such as broadband and Internet access. They point to a number of provisions contained in the Universal Service section of the 1996 act to support their claim. Universal service principles contained in Section 254(b)(2) state that "Access to advanced telecommunications services should be provided to all regions of the Nation." The subsequent principle (b)(3) calls for consumers in all regions of the nation including "low-income" and those in "rural, insular, and high cost areas" to have access to telecommunications and information services including "advanced services" at a comparable level and a comparable rate charged for similar services in urban areas. Such provisions, they state, dictate that the FCC expand its universal service definition.

The 1996 act does take into consideration the changing nature of the telecommunications sector and allows for the universal service definition to be modified if future conditions warrant. Section 254(c)of the act states that "universal service is an evolving level of telecommunications services" and the FCC is tasked with "periodically" reevaluating this definition "taking into account advances in telecommunications and information technologies and services." Furthermore, the Joint Board is given specific authority to recommend "from time to time" to the FCC modification in the definition of the services to be included for federal universal service support. The Joint Board, on November 19, 2007, concluded such an inquiry and recommended that the FCC change the mix of services eligible for universal service support. The Joint Board recommended, among other things, that "the universal availability of broadband Internet services" be included in the nation's communications goals and hence be supported by federal universal service funds.[49]

This debate was put to rest when provisions contained in the American Recovery and Reinvestment Act of 2009 (ARRA) called for the FCC to develop, and submit to Congress, a national broadband plan to ensure that every American has "access to broadband capability."[50] The FCC in its national broadband plan, *Connecting America: the National Broadband Plan,* recommended that access to and adoption of broadband be a national goal. Furthermore the national broadband plan proposed that the Universal Service Fund be restructured to become a vehicle to help reach this goal. The FCC, in an October 2011 decision, adopted an Order that calls for the USF to be transformed, in stages, over a multi-year period, from a mechanism to support voice telephone service to one that supports the deployment, adoption, and utilization of both fixed and mobile broadband.

This transformation includes the phase out of the USF's legacy High Cost Program and the creation of a new fund, the Connect America Fund, to replace it.[51]

LEGISLATION IN THE 110TH CONGRESS

In the 110th Congress, legislation was enacted to provide financial assistance for broadband deployment. Of particular note is the reauthorization of the Rural Utilities Service (RUS) broadband loan program, which was enacted as part of the 2008 farm bill (P.L. 110-234). In addition to reauthorizing and reforming the RUS broadband loan program, P.L. 110-234 contains provisions establishing a National Center for Rural Telecommunications Assessment and requiring the FCC and RUS to formulate a comprehensive rural broadband strategy.

The Broadband Data Improvement Act (P.L. 110-385) was enacted by the 110th Congress and required the FCC to collect demographic information on unserved areas, data comparing broadband service with 75 communities in at least 25 nations abroad, and data on consumer use of broadband. The act also directed the Census Bureau to collect broadband data, the Government Accountability Office to study broadband data metrics and standards, and the Department of Commerce to provide grants supporting state broadband initiatives.

Meanwhile, the America COMPETES Act (H.R. 2272) was enacted (P.L. 110-69) and contained a provision authorizing the National Science Foundation (NSF) to provide grants for basic research in advanced information and communications technologies. Areas of research included affordable broadband access, including wireless technologies. P.L. 110-69 also directs NSF to develop a plan that describes the current status of broadband access for scientific research purposes.

LEGISLATION IN THE 111TH CONGRESS

In the 111th Congress, legislation was enacted that sought to provide financial assistance for broadband deployment. Of particular note, provisions in the American Recovery and Reinvestment Act of 2009 (P.L. 111-5) provided grants and loans to support broadband access and adoption in unserved and underserved areas.

P.L. 111-5: The American Recovery and Reinvestment Act of 2009

On February 17, 2009, President Obama signed P.L. 111-5, the American Recovery and Reinvestment Act (ARRA). Broadband provisions of the ARRA provided a total of *$7.2 billion*, for broadband grants, loans, and loan/grant combinations.

The total consisted of $4.7 billion to NTIA/DOC for a newly established Broadband Technology Opportunities Program (grants) and $2.5 billion to the RUS/USDA Broadband Initiatives Program (grants, loans, and grant/loan combinations).[52]

Regarding the $2.5 billion to RUS/USDA broadband programs, the ARRA specified that at least 75% of the area to be served by a project receiving funds shall be in a rural area without sufficient access to high speed broadband service to facilitate economic development, as determined by the Secretary of Agriculture.

Priority was given to projects that provide service to the most rural residents that do not have access to broadband services. Priority was also given to borrowers and former borrowers of rural telephone loans.

Of the $4.7 billion appropriated to NTIA:

- $4.35 billion was directed to a competitive broadband grant program, of which not less than $200 million shall be available for competitive grants for expanding public computer center capacity (including at community colleges and public libraries); not less than $250 million to encourage sustainable adoption of broadband service; and $10 million transferred to the Department of Commerce Office of Inspector General for audits and oversight; and

- $350 million was directed for funding the Broadband Data Improvement Act (P.L. 110-385) and for the purpose of developing and maintaining a broadband inventory map, which shall be made accessible to the public no later than two years after enactment. Funds deemed necessary and appropriate by the Secretary of Commerce may be transferred to the FCC for the purposes of developing a national broadband plan, which shall be completed one year after enactment.

Final BTOP and BIP program awards were announced by September 30, 2010.

Other Enacted Broadband Legislation in the 111th Congress

P.L. 111-8 (H.R. 1105). Omnibus Appropriations Act, 2009. Appropriates to RUS/USDA $15.619 million to support a loan level of $400.487 million for the Rural Broadband Access Loan and Loan Guarantee Program, and $13.406 million for the Community Connect Grant Program.

To the FCC, designates not less than $3 million to establish and administer a State Broadband Data and Development matching grants program for state-level broadband demand aggregation activities and creation of geographic inventory maps of broadband service to identify gaps in service and provide a baseline assessment of statewide broadband deployment. Passed House February 25, 2009. Passed Senate March 10, 2009. Signed by President, March 12, 2009.

P.L. 111-32 (H.R. 2346). Supplemental Appropriations Act, 2009. Provides not less than $3 million to the FCC to develop a national broadband plan pursuant to the American Recovery and Reinvestment Act of 2009. Introduced May 12, 2009; referred to Committee on Appropriations. Passed House May 14, 2009; passed Senate May 21, 2009. Signed by President, June 24, 2009.

P.L. 111-80 (H.R. 2997). Agriculture, Rural Development, Food and Drug Administration, and Related Agencies Appropriations Act, 2010. For Rural Utilities Service, U.S. Department of Agriculture, provides $28.96 million to support a loan level of $400 million for the broadband loan program, and $17.97 million for broadband community connect grants. Introduced June 23, 2009; referred to Committee on Appropriations. Reported by Committee on Appropriations June 23, 2009. Passed House July 9, 2009. Passed Senate

August 4, 2009. Conference Report (H.Rept. 111-279) printed September 30, 2009. Signed by President October 21, 2009.

LEGISLATION IN THE 112TH CONGRESS

The 112th Congress examined the efficacy of federal broadband assistance programs and how they may fit into the context of a national broadband policy.

P.L. 112-10 (H.R. 1473). Department of Defense and Full-Year Continuing Appropriations Act, 2011. Rescinds existing unobligated past-year funding for the Rural Broadband Access Loan and Loan Guarantee Program and the Community Connect Grants at the Rural Utilities Service. For FY2011, appropriates $22.3 million to the Rural Broadband Access Loan and Loan Guarantee Program for the cost of broadband loans, and $13.4 million to Community Connect Grants. Signed by President, April 15, 2011.

P.L. 112-55 (H.R. 2112). Consolidated and Further Continuing Appropriations Act, 2012. Provides FY2012 appropriations for Rural Utilities Service broadband loan program and broadband community connect grants: $6 million for the broadband loan program (subsidizing a loan level of $212 million) and $10.372 million for Community Connect grants. Introduced June 3, 2011; referred to Committee on Appropriations. Reported by Committee on Appropriations June 3, 2011 (H.Rept. 112-101). Passed House June 16, 2011. Reported by Senate Appropriations Committee September 7, 2011 (S.Rept. 112-73). Signed by President, November 18, 2011.

H.R. 1083 (Owens). Rural Broadband Initiative Act. Establishes an Office of Rural Broadband Initiatives in the Department of Agriculture which would administer the RUS broadband loan and grant programs, and would develop a comprehensive rural broadband strategy. Introduced March 15, 2011; referred to Committee on Agriculture and in addition to the Committee on Energy and Commerce.

H.R. 1343 (Bass). To return unused or reclaimed funds made available for broadband awards in the American Recovery and Reinvestment Act of 2009 to the Treasury of the United States. Introduced April 4, 2011; referred to Committee on Energy and Commerce and to Committee on Agriculture. Reported (amended) by the Committee on Energy and Commerce (H.Rept. 112-228) on September 29, 2011. Passed House October 5, 2011. Referred to Senate Committee on Commerce, Science and Transportation October 6, 2011.

H.R. 1695 (Eshoo). Broadband Conduit Deployment Act of 2011. Directs the Secretary of Transportation to require that broadband conduit be installed as part of certain highway construction projects. Introduced May 3, 2011; referred to Committee on Transportation and Infrastructure.

H.R. 2163 (Matsui). Broadband Affordability Act of 2011. Amends the Communications Act of 1934 to establish a Lifeline Assistance Program for universal broadband adoption. Introduced June 14, 2011; referred to Committee on Energy and Commerce.

H.R. 5973 (Kingston). Agriculture, Rural Development, Food and Drug Administration, and Related Agencies Appropriations Act, 2013. For Rural Utilities Service, U.S. Department of Agriculture, provides $2 million to support a loan level of $21 million for the broadband loan program, and $10 million for broadband community connect grants. Introduced June 20, 2012; referred to Committee on Appropriations. Reported by Committee on Appropriations June 20, 2012.

H.R. 6083 (Lucas). Federal Agriculture Reform and Risk Management Act of 2012. Reauthorizes rural broadband loan program at $25 million per year through FY2017. Introduced July 9, 2012; referred to Committee on Agriculture. Ordered to be reported by committee July 11, 2012.

S. 257 (Landrieu). Small Business Broadband and Emerging Information Technology Enhancement Act of 2011. Seeks to improve certain programs of the Small Business Administration to better assist small business customers in accessing broadband technology. Introduced February 2, 2011; referred to Committee on Small Business and Entrepreneurship.

S. 1659 (Ayotte). To return unused or reclaimed funds made available for broadband awards in the American Recovery and Reinvestment Act of 2009 to the Treasury of the United States. Introduced October 5, 2011; referred to Committee on Commerce, Science and Transportation.

S. 1939 (Klobuchar). Broadband Conduit Deployment Act of 2011. Directs the Secretary of Transportation to require that broadband conduit be installed as part of certain highway construction projects. Introduced December 1, 2011; referred to Committee on Environment and Public Works.

S. 2298 (Brown of Ohio). Connecting Rural America Act. Amends the Rural Electrification Act of 1936 to improve the program of access to broadband telecommunications services in rural areas. Introduced April 18, 2002; referred to Committee on Agriculture, Nutrition, and Forestry.

S. 2375 (Kohl). Agriculture, Rural Development, Food and Drug Administration, and Related Agencies Appropriations Act, 2013. For Rural Utilities Service, U.S. Department of Agriculture, provides $6 million to support a loan level of $63 million for the broadband loan program, and $10 million for broadband community connect grants. Introduced April 26, 2012; referred to Committee on Appropriations. Reported by Committee on Appropriations April 26, 2012.

S. 3240 (Stabenow). Agriculture Reform, Food, and Jobs Act of 2012. Authorizes broadband loan and grant program at $50 million per year through FY2017. Introduced May 24, 2012; referred to Committee on Agriculture, Nutrition and Forestry. Reported to Senate May 24, 2012. Passed Senate (amended) June 21, 2012.

S. 3439 (Snowe). Federal Wi-Net Act. Directs the Administrator of General Services to install Wi-Fi hotspots and wireless neutral host systems in all federal buildings in order to improve in-building wireless communications coverage and commercial network capacity by offloading wireless traffic onto wireline broadband networks. Introduced July 25, 2012; referred to Committee on Environment and Public Works.

LEGISLATION IN THE 113TH CONGRESS

The following is a listing of broadband legislation directly related to the issue of federal assistance for broadband deployment in unserved areas.[53]

P.L. 113-6 (H.R. 933). Consolidated and Further Continuing Appropriations Act, 2013.Funds the broadband loan program at $4 million (supporting a loan level of approximately $42 million) and the Community Connect grant program at $10.372 million. Signed by President March 26, 2013.

H.R. 2163 (Matsui). Broadband Adoption Act of 2013. Amends the Communications Act of 1934 to reform and modernize the Universal Service Fund Lifeline Assistance Program. Introduced April 26, 2013; referred to Committee on Energy and Commerce.

H.R. 1947 (Lucas). Federal Agriculture Reform and Risk Management Act of 2013. Section 6105 would reauthorize the broadband loan and loan guarantee program through FY2018 at the current level of $25 million per year. Introduced May 13, 2013; reported by Committees on Agriculture and Judiciary.

H.R. 1639 (Gibson). Amends the Rural Electrification Act of 1936 to authorize loan/grant combinations under RUS broadband program. Introduced April 18, 2013; referred to Committee on Agriculture and Committee on Energy and Commerce.

H.R. 2410 (Aderholt). Agriculture, Rural Development, Food and Drug Administration, and Related Agencies Appropriations Act, 2014. Provides $5.5 million to subsidize a loan level of $42.146 million for the broadband loan program, and $10.111 million for the Community Connect grant program. Introduced June 18, 2013; reported by House Committee on Appropriations (H.Rept. 113-116).

S. 954 (Stabenow). Agriculture Reform, Food, and Jobs Act of 2013. Includes the establishment of a new grant program in combination with the existing loan and loan guarantee program authorization, which is extended at $50 million per year through FY2018. Passed by Senate June 10, 2013.

CONCLUDING OBSERVATIONS

To the extent that the 113th Congress may consider various options for encouraging broadband deployment and adoption, a key issue is how to strike a balance between providing federal assistance for unserved and underserved areas where the private sector may not be providing acceptable levels of broadband service, while at the same time minimizing any deleterious effects that government intervention in the marketplace may have on competition and private sector investment.

In addition to loans, loan guarantees, and grants for broadband infrastructure deployment, a wide array of policy instruments are available to policymakers, including universal service reform, tax incentives to encourage private sector deployment, broadband bonds, demand-side incentives (such as assistance to low income families for purchasing computers), regulatory and deregulatory measures, and spectrum policy to spur roll-out of wireless broadband services. In assessing federal incentives for broadband deployment, the 113th Congress may consider the appropriate mix of broadband deployment incentives to create jobs in the short and long term, the extent to which incentives should target next-generation broadband technologies, the extent to which "underserved" areas with existing broadband providers should receive federal assistance, and whether broadband stimulus projects are being efficiently managed and how they may fit into the context of overall goals for a national broadband policy.

Table 4. Selected Federal Domestic Assistance Programs Related to Broadband and Telecommunications Development

Program	Agency	Description	Funding Amount (est. FY2012 unless otherwise noted)	Web Links
		Broadband Infrastructure Deployment Programs		
Broadband Technology Opportunities Program (BTOP)	National Telecommunications and Information Administration, Dept. of Commerce	Provides competitive grants to public and private sector entities in order to provide broadband access in unserved and underserved areas; provide broadband support and services to strategic institutions; improve broadband access by public safety agencies; and stimulate broadband demand, economic growth, and job creation.	$4 billion (ARRA, P.L. 111-5) (2009)	http://www.ntia.doc.gov/ broadbandgrants/
Broadband Initiatives Program (BIP)	Rural Utilities Service, U.S. Dept. of Agriculture	Provides competitive grants, loans, and loan/grant combinations to public and private sector entities in order to provide broadband access in unserved and underserved rural areas.	$2.5 billion for the cost of loans, grants, and loan/grant combinations (ARRA, P.L. 111-5) (2009)	http://www.rurdev.usda.gov/utp_bip.html
Rural Broadband Access Loan and Loan Guarantee Program	Rural Utilities Service, U.S. Dept. of Agriculture	Provides loan and loan guarantees for facilities and equipment providing broadband service in rural communities	$169 million for cost of money loans	http://www.rurdev.usda.gov/utp_farmbill.html
Community Connect Broadband Grants	Rural Utilities Service, U.S. Dept. of Agriculture	Provides grants to applicants proposing to provide broadband service on a "community-oriented connectivity" basis to rural communities of under 20,000 inhabitants.	$10 million	http://www.rurdev.usda.gov/utp_commconnect.html

Table 4. (Continued)

Program	Agency	Description	Funding Amount (est. FY2012 unless otherwise noted)	Web Links
	Telecommunications Broadband Infrastructure Deployment Programs			
Telecommunications Infrastructure Loan Program	*Rural Utilities Service, U.S. Dept. of Agriculture*	Provides long-term direct and guaranteed loans to qualified organizations for the purpose of financing the improvement, expansion, construction, acquisition, and operation of telephone lines, facilities, or systems to furnish and improve telecommunications service in rural areas. All facilities financed must be capable of supporting broadband services.	$145 million for hardship loans; $250 million for cost of money loans; and $295 million for FFB Treasury loans	http://www.rurdev.usda.gov/ utp_infrastructure.html
Distance Learning and Telemedicine Loans and Grants	*Rural Utilities Service, U.S. Dept. of Agriculture*	Provides seed money to rural community facilities (e.g., schools, libraries, hospitals) for advanced telecommunications systems that can provide health care and educational benefits to rural areas	$21 million	http://www.rurdev.usda.gov/ UTP_DLT.html
Universal Service High Cost Program[a]	*Federal Communications Commission*	Provides funding to eligible telecommunications carriers to help pay for telecommunications services in high-cost, rural, and insular areas so that prices charged to customers are reasonably comparable across all regions of the nation.	$4.5 billion (annually through 2017)	http://www.usac.org/hc/
Universal Service Schools and Libraries	*Federal Communications*	Provides discounts for affordable telecommunications and Internet access	$2.25 billion (annually adjusted for inflation)	http://www.universalservice.org/sl/

Program	Agency	Description	Funding Amount (est. FY2012 unless otherwise noted)	Web Links
Program (i.e., E-rate)	*Commission*	services to ensure that schools and libraries have access to affordable telecommunications and information services.		
Universal Service Rural Health Care Pilot Program[b]	*Federal Communications Commission*	Provides funds to cover 85% of the cost of constructing statewide or regional broadband telehealth networks and of connecting those projects to dedicated nationwide broadband telehealth networks and the public Internet.	$418 million (annually)	http://www.usac.org/rhc-pilot-program/
Appalachian Area Development Program	*Appalachian Regional Commission*	Project grants to support self-sustaining economic development in the region's most distressed counties and areas. Includes funds for a Telecommunications Initiative involving projects that enable communities to capitalize on broadband access.	$56 million	http://www.arc.gov/index.do?nodeId= 21
States' Economic Development Assistance Program	*Delta Regional Authority*	Grants for self-sustaining economic development projects of eight states in Mississippi Delta region.	$11 million (2011)	http://grants.dra.gov/
Investments for Public Works and Economic Development Facilities	*Economic Development Administration, Dept. of Commerce*	Provides funding for construction of infrastructure in areas that are not attractive to private investment; most funding is for water and sewer infrastructure but some has been designated for telecommunications and broadband projects.	$112 million	http://www.eda.gov/

Table 4. (Continued)

Program	Agency	Description	Funding Amount (est. FY2012 unless otherwise noted)	Web Links
Library Services and Technology Act Grants to States	*Institute of Museum and Library Services, National Foundation on the Arts and the Humanities*	Provides funds for a wide range of library services including installation of fiber and wireless networks that provide access to library resources and services.	$156 million	http://www.imls.gov/programs/programs.shtm
Native American Library Services	*Institute of Museum and Library Services, National Foundation on the Arts and the Humanities*	Grants to support library services including electronically linking libraries to networks.	$4 million	http://www.imls.gov/applicants/grants/nativeAmerican.shtm
Choice Neighborhood Implementation Grants	*Office of the Assistant Secretary for Public and Indian Housing and Office of Multifamily Housing Programs, Dept. of Housing and Urban Development*	Helps communities transform neighborhoods by revitalizing severely distressed public and/or assisted housing. Grantees may use funds to provide unit-based broadband Internet connectivity.	$110 million	http://www.hud.gov/cn/
Special Education—Technology and Media Services for	*Office of Special Education and Rehabilitative Services, Dept. of*	Supports development and application of technology and education media activities for disabled children and adults	$30 million	http://www.ed.gov/about/offices/list/osers/index.html?src=mr/

Program	Agency	Description	Funding Amount (est. FY2012 unless otherwise noted)	Web Links
Individuals with Disabilities	*Education*			
Telehealth Network Grants	*Health Resources and Services Administration, Department of Health and Human Services*	Grants to develop sustainable telehealth programs and networks in rural and frontier areas, and in medically unserved areas and populations.	$6 million	http://www.hrsa.gov/telehealth/
Telehealth Resource Center Grant Program	*Health Resources and Services Administration, Department of Health and Human Services*	Provides grants that support establishment and development of telehealth resource centers to assist health care providers in the development of telehealth services, including decisions regarding the purchase of advanced telecommunications services.	$4 million	http://www.hrsa.gov/telehealth/
Licensure Portability Grant Program	*Health Resources and Services Administration, Department of Health and Human Services*	Provides support for state professional licensing boards to develop and implement state policies that will reduce statutory and regulatory barriers to telemedicine.	$0.35 million	http://www.hrsa.gov/telehealth/
NLM Extramural Programs	*National Library of Medicine, National Institutes of Health,*	Provides funds to train professional personnel; strengthen library and information services; facilitate access to and delivery of	$62 million	http://www.nlm.nih.gov/ep/extramural.html

Table 4. (Continued)

Program	Agency	Description	Funding Amount (est. FY2012 unless otherwise noted)	Web Links
	Department of Health and Human Services	health science information; plan and develop advanced information networks; support certain kinds of biomedical publications; and conduct research in medical informatics and related sciences.		
National Environmental Information Exchange Network Grant Program	*Environmental Protection Agency*	Provides funding to states, territories, and federally recognized Indian Tribes to support the development of an Environmental Information Exchange Network, including broadband infrastructure.	$10 million	http://epa.gov/exchange network/grants/

Source: Compiled by CRS from FY2013 budget documents, the Catalog of Federal Domestic Assistance, and grants.

[a] The High Cost program will be phased out, and replaced in stages by the Connect America Fund. The program provides funding to eligible service providers to support the provision of affordable voice and broadband services, both fixed and mobile.

[b] The Rural Health Care program will be replaced by the Healthcare Connect Fund.

End Notes

[1] The term "digital divide" can also refer to international disparities in access to communications and information technology. This report focuses on domestic issues only.

[2] FCC, Internet Access Services: Status as of June 30, 2012, released May 2013, p. 17. Available at http://transition.fcc.gov/Daily_Releases/Daily_Business/2013/db0520/DOC-321076A1.pdf.

[3] Ibid.

[4] Initially, and for many years following, the FCC defined broadband (or more specifically "high-speed lines") as over 200 kilobits per second (kbps) in at least one direction, which was roughly four times the speed of conventional dialup Internet access. In recent years, the 200 kbps threshold was considered too low, and on March 19, 2008, the FCC adopted a report and order (FCC 08-89) establishing new categories of broadband speed tiers for data collection purposes. Specifically, 200 kbps to 768 kbps is considered "first generation," 768 kbps to 1.5 Mbps is "basic broadband tier 1," and increasingly higher speed tiers are broadband tiers 2 through 7 (tier seven is greater than or equal to 100 Mbps in any one direction). Tiers can change as technology advances.

[5] NTIA, U.S. Broadband Availability: June 2010 – June 2012, May 2013, p.5.available at http://www.ntia.doc.gov/files/ ntia/publications/usbb_avail_report_05102013.pdf.

[6] Federal Communications Commission, Eighth Broadband Progress Report, FCC 12-90, released August 21, 2012, p. 29, available at http://transition.fcc.gov/Daily_Releases/Daily_Business/2012/db0827/FCC-12-90A1.pdf.

[7] U.S. Department of Commerce, National Telecommunications and Information Administration, Blog, "Household Broadband Adoption Climbs to 72.4 Percent," June 6, 2013, available at http://www.ntia.doc.gov/blog/2013/householdbroadband-adoption-climbs-724-percent.

[8] Eighth Broadband Progress Report, p. 54.

[9] U.S. Department of Commerce, National Telecommunications and Information Administration, Exploring the Digital Nation: America's Emerging Online Experience, June 2013, p. 26, available at http://www.ntia.doc.gov/files/ntia/ publications/exploring_the_digital_nation_-_americas_emerging_online_experience.pdf.

[10] Smith, Aaron, Pew Internet & American Life Project, Home Broadband 2010, August 11, 2010, available at http://www.pewinternet.org/~/media//Files/Reports/2010/ Home%20broadband%202010.pdf.

[11] For more information on rural broadband and broadband programs at the Rural Utilities Service, see CRS Report RL33816, Broadband Loan and Grant Programs in the USDA's Rural Utilities Service, by Lennard G. Kruger.

[12] See for example: National Exchange Carrier Association (NECA), Trends 2006: Making Progress With Broadband, 2006, 26 p. Available at http://www.neca.org/media/trends_brochure_website.pdf.

[13] Federal Communications Commission, Eighth Broadband Progress Report, FCC 12-90, released August 21, 2012, p. 5, available at http://transition.fcc.gov/Daily_Releases/Daily_Business/2012/db0827/FCC-12-90A1.pdf.

[14] NTIA, National Broadband Map, Broadband Statistics Report: Broadband Availability in Urban vs. Rural Areas, January 2013, p. 7, available at http://www.broadbandmap.gov/download/ Broadband%20Availability%20in%20Rural%20vs%20Urban%20Areas.pdf. Also see NTIA, U.S. Broadband Availability: June 2010–June 2012, May 2013, p. 10-11, available at http://www.ntia.doc.gov/files/ntia/publications/ usbb_avail_report_ 05102013.pdf.

[15] U.S. Department of Commerce, National Telecommunications and Information Administration, Exploring the Digital Nation: America's Emerging Online Experience, June 2013, p. 26, available at http://www.ntia.doc.gov/files/ ntia/publications/exploring_the_digital_nation_-_americas_emerging_online_experience.pdf.

[16] The terrain of rural areas can also be a hindrance to broadband deployment because it is more expensive to deploy broadband technologies in a mountainous or heavily forested area. An additional added cost factor for remote areas can be the expense of "backhaul" (e.g., the "middle mile") which refers to the installation of a dedicated line which transmits a signal to and from an Internet backbone which is typically located in or near an urban area.

[17] Gillett, Sharon E., Massachusetts Institute of Technology, Measuring Broadband's Economic Impact, report prepared for the Economic Development Administration, U.S. Department of Commerce, February 28, 2006, p. 4.

[18] Crandall, Robert, William Lehr, and Robert Litan, The Effects of Broadband Deployment on Output and Employment: A Cross-sectional Analysis of U.S. Data, June 2007, 20 pp. Available at http://www3.brookings.edu/ views/papers/crandall/200706litan.pdf.

[19] Mark Dutz, Jonathan Orszag, and Robert Willig, The Substantial Consumer Benefits of Broadband Connectivity for U.S. Households, Internet Innovation Alliance, July 2009, p. 4, http://internetinnovation.org/files/special-reports/ CONSUMER_BENEFITS_OF_BROADBAND.pdf.

[20] OECD, OECD Broadband Portal. Available at http://www.oecd.org/sti/ict/broadband.

[21] OECD, Directorate for Science, Technology and Industry, The Development of Broadband Access in OECD Countries, October 29, 2001, 63 pp. For a comparison of government broadband policies, also see OECD, Directorate for Science, Technology and Industry, Broadband Infrastructure Deployment: The Role of Government Assistance, May 22, 2002, 42 pp.

[22] See Turner, Derek S., Free Press, Broadband Reality Check II: The Truth Behind America's Digital Divide, August 2006, pp. 8-11. Available at http://www.freepress.net/files/bbrc2-final.pdf; and Turner, Derek S., Free Press, 'Shooting the Messenger' Myth vs. Reality: U.S. Broadband Policy and International Broadband Rankings, July 2007, 25 pp., available at http://www.freepress.net/files/shooting_the_messenger.pdf.

[23] National Telecommunications and Information Administration, Fact Sheet: United States Maintains Information and Communication Technology (ICT) Leadership and Economic Strength, at http://www.ntia.doc.gov/ntiahome/press/ 2007/ICTleader_042407.html.

[24] See Wallsten, Scott, Progress and Freedom Foundation, Towards Effective U.S. Broadband Policies, May 2007, 19 pp. Available at http://www.pff.org/issues-pubs/pops/pop14.7 usbroadbandpolicy.pdf. Also see Ford, George, Phoenix Center, The Broadband Performance Index: What Really Drives Broadband Adoption Across the OECD?, Phoenix Center Policy Paper Number 33, May 2008, 27 pp.; available at http://www.phoenix-center.org/pcpp/PCPP33Final.pdf.

[25] See price and services and speed data on OECD Broadband Portal, available at http://www.oecd.org/sti/ict/broadband; see also Federal Communications Commission, Third Annual International Broadband Data Report, IB Docket No. 10-171, DA 12-1334, August 21, 2012, available at http://hraunfoss.fcc.gov/edocs_public/attachmatch/DA12-1334A1.pdf.

[26] Federal Communications Commission, Notice of Inquiry, "Concerning the Deployment of Advanced Telecommunications Capability to All Americans in a Reasonable and Timely Fashion, and possible Steps to Accelerate Such Deployment Pursuant to Section 706 of the Telecommunications Act of 1996," FCC 04-55, March 17, 2004, p. 6.

[27] FCC News Release, FCC Improves Data Collection to Monitor Nationwide Broadband Rollout, November 9, 2004. Available at http://hraunfoss.fcc.gov/edocs_public/attachmatch/DOC-254115A1.pdf.

[28] Federal Communications Commission, Notice Proposed Rulemaking, "Development of Nationwide Broadband Data to Evaluate Reasonable and Timely Deployment of Advanced Services to All Americans, Improvement of Wireless Broadband Subscribership Data, and Development of Data on Interconnected Voice Over Internet Protocol (VoIP) Subscribership," WC Docket No. 07-38, FCC 07-17, released April 16, 2007, 56 pp.

[29] FCC, News Release, "FCC Expands, Improves Broadband Data Collection," March 19, 2008. Available at http://hraunfoss.fcc.gov/edocs_public/attachmatch/DOC-280909A1.pdf.

[30] For more information on the national broadband mapping program and the State Broadband Data and Development Program, see http://www.ntia.doc.gov/broadband grants/broadbandmapping.html.

[31] Eighth Broadband Progress Report, p. 59-60.

[32] Ibid., p.171, 177.

[33] Available at http://www.broadband.gov/plan/For more information on the National Broadband Plan, see CRS Report R41324, The National Broadband Plan, by Lennard G. Kruger et al..

[34] Federal Communications Commission, Connecting America: The National Broadband Plan, March 17, 2010, p. 9.

[35] Ibid., p. 11.

[36] Ibid.

[37] Ibid.

[38] See for example, Office of Science and Technology Policy and National Economic Council, The White House, Four Years of Broadband Growth, June 2013, 26 pages, available at http://www.whitehouse.gov/sites/default/files/ broadband_report_final.pdf.

[39] See CRS Report R40436, Broadband Infrastructure Programs in the American Recovery and Reinvestment Act, by Lennard G. Kruger.

[40] For more information on these programs, see CRS Report RL33816, Broadband Loan and Grant Programs in the USDA's Rural Utilities Service, by Lennard G. Kruger.

[41] Ibid.

[42] See CRS Report R42524, Rural Broadband: The Roles of the Rural Utilities Service and the Universal Service Fund, by Angele A. Gilroy and Lennard G. Kruger.

[43] The section on universal service was prepared by Angele Gilroy, Specialist in Telecommunications, Resources, Science and Industry Division. For more information on universal service, see CRS Report RL33979, Universal Service Fund: Background and Options for Reform, by Angele A. Gilroy.

[44] Communications Act of 1934, As Amended, Title I §1 [47 U.S.C. 151].

[45] Many states participate in or have programs that mirror FCC universal service mechanisms to help promote universal service goals within their states.

[46] For additional information on this program, including funding commitments, see the E-rate website: http://www.universalservice.org/sl/.

[47] For additional information on this program, including funding commitments, see the RHCD website: http://www.universalservice.org/rhc/.

[48] For more details on the USF rural health care support mechanism and the newly established Healthcare Connect Fund see In the Matter of Rural Health Care Support Mechanism, WC Docket No. 02-60, Federal Communications Commission, adopted December 12, 2012. Available at http://hraunfoss.fcc.gov/edocs_public/attachmatch/FCC-12- 150A1.pdf.

49 The Joint Board recommended that the definition of those services that qualify for universal service support be expanded and that the nation's communications goals include the universal availability of: mobility services (i.e., wireless voice); broadband Internet services; and voice services at affordable and comparable rates for all rural and non-rural areas. For a copy of this recommendation see http://hraunfoss.fcc.gov/edocs_public/attachmatch/FCC-07J4A1.pdf.

50 American Recovery and Reinvestment Act of 2009, P.L. 111-5, sec. 6001 (k)(2)(D).

51 For a detailed discussion of this Order and USF transition see CRS Report R42524, Rural Broadband: The Roles of the Rural Utilities Service and the Universal Service Fund, by Angele A. Gilroy and Lennard G. Kruger.

52 For information on existing broadband programs at RUS, see CRS Report RL33816, Broadband Loan and Grant Programs in the USDA's Rural Utilities Service, by Lennard G. Kruger.

53 For information on public safety wireless broadband legislation, see CRS Report R41842, Funding Emergency Communications: Technology and Policy Considerations, by Linda K. Moore.

INDEX

C

G

H

I

Q

R

S

V

W